BORROWED TIME

Also available in the Bloomsbury Sigma series:

P53: The Gene that Cracked the Cancer Code by Sue Armstrong
Spirals in Time by Helen Scales
A is for Arsenic by Kathryn Harkup
Breaking the Chains of Gravity by Amy Shira Teitel
Herding Hemingway's Cats by Kat Arney
Sorting the Beef from the Bull by Richard Evershed and Nicola Temple
Death on Earth by Jules Howard
The Tyrannosaur Chronicles by David Hone
Soccermatics by David Sumpter
Big Data by Timandra Harkness
Goldilocks and the Water Bears by Louisa Preston
Science and the City by Laurie Winkless
Bring Back the King by Helen Pilcher
Furry Logic by Matin Durrani and Liz Kalaugher
Built on Bones by Brenna Hassett
My European Family by Karin Bojs
4th Rock from the Sun by Nicky Jenner
Patient H69 by Vanessa Potter
Catching Breath by Kathryn Lougheed
PIG/PORK by Pía Spry-Marqués
The Planet Factory by Elizabeth Tasker
Wonders Beyond Numbers by Johnny Ball
Immune by Catherine Carver
I, Mammal by Liam Drew
Reinventing the Wheel by Bronwen and Francis Percival
Making the Monster by Kathryn Harkup
Best Before by Nicola Temple
Catching Stardust by Natalie Starkey
Seeds of Science by Mark Lynas
Outnumbered by David Sumpter
Eye of the Shoal by Helen Scales
Nodding Off by Alice Gregory
The Science of Sin by Jack Lewis
The Edge of Memory by Patrick Nunn
Turned On by Kate Devlin

BORROWED TIME

The Science of How and Why We Age

Sue Armstrong

B L O O M S B U R Y   S I G M A
LONDON • OXFORD • NEW YORK • NEW DELHI • SYDNEY

BLOOMSBURY SIGMA
Bloomsbury Publishing Plc
50 Bedford Square, London, WC1B 3DP, UK

BLOOMSBURY, BLOOMSBURY SIGMA and the Bloomsbury Sigma logo
are trademarks of Bloomsbury Publishing Plc

First published in the United Kingdom in 2019

A catalogue record for this book is available from the British Library

Library of Congress Cataloguing-in-Publication data has been applied for

ISBN: HB: 978-1-4729-3606-6; TPB: 978-1-4729-3609-7;
eBook: 978-1-4729-3607-3

2 4 6 8 10 9 7 5 3 1

Typeset by Deanta Global Publishing Services, Chennai, India
Printed and bound in Great Britain by CPI Group (UK) Ltd, Croydon CR0 4YY

Bloomsbury Sigma, Book Forty-one

We all live most of our lives on borrowed time, time that our stepmother nature never intended us to have.

Bryan Appleyard

For my sisters, Jane and Julie, who have shared this whole journey with me so far; and for Fred, who joined me halfway

Contents

Preface

Consider this: the Greenland shark can live for more than 400 years and appears to remain physically fit and fertile to the end. Or this: there is a type of jellyfish that swims the Mediterranean and the seas around Japan whose individuals are able to revert to the larval state and regrow to adulthood countless times. In other words, it is biologically immortal. So too the hydra, familiar to many of us from our first biology lessons looking at drops of pond water under a microscope: its body is composed entirely of immortal stem cells, and a whole new hydra can be regenerated from any little piece that gets chopped off. These last two creatures seem to possess the gift of eternal youth and vigour – and never die of old age, as far as anyone knows for sure.

The question of exactly how and why organisms – most especially us – age has teased scientists for centuries, yet there is still no agreement. There are a myriad competing theories, from the built-in obsolescence of the 'disposable soma' theory (which basically proposes that Nature doesn't have much use for us once we've passed our reproductive years and hasn't invested in good enough repair and maintenance systems to keep us going indefinitely) and the idea that ageing is wear and tear like the rusting of a car or the weathering of a canvas tent, to the ticking clock of the shortening telomeres that measure out the lifespan of our dividing cells, and the idea that ageing and death are genetically programmed and controlled. A growing number of respected scientists even believe that ageing is a disease, and it can be treated. Some go so far as to suggest ageing can be 'cured', so that we too could potentially live forever.

I found this last idea – the quest for 'immortal life' – so exasperatingly narcissistic that I contemplated giving up on this book near the start. But since I had already booked a flight to California (where else?) and made plans to meet a bunch of scientists when the big doubts struck, I decided to go anyway, enjoy the trip and make a decision when I had talked to some leading people in ageing research, or gerontology as it's called. One of my first interviewees, when asked how he felt about certain colleagues who claim we are on the brink of being able to extend the human lifespan to 150, 500, 1,000 years and more, answered, 'I'd say what are they smoking?' As he headed off to another meeting at the end of our interview, he quipped, 'Send me a postcard when you get to those promised shores!'

A good laugh helped restore my faith in my project and I decided happily to carry on. Over the course of my research I have met some fascinating people, engaged in enthralling debate, and been forced to confront my own prejudices, for I, like most of us I suspect, have taken ageing for granted as an inevitable process to be accepted and endured, if not welcomed. But the fact is that the biggest single risk factor for a host of conditions – from stiffening joints, thinning bones and waning energy to heart failure, cancer, stroke, dementia and the steady loss of hearing and eyesight – is old age.

This irrefutable fact makes the quest to tease out the details of why and how our bodies degenerate and whether we can intervene in the process an eminently worthwhile endeavour, for the ageing of the world population is up there with climate change as one of the biggest challenges of the twenty-first century. It has implications for every aspect of society, from how our economies are managed and goods and services provided to cater for everyone's needs, to our working lives, politics, the relationships between the generations and the dynamics of family life.

As we have progressively won the battle against the infectious and parasitic diseases that were the leading killers of generations past, life expectancy at birth for the global population as a whole has gone up from just 48 years in 1955 to more than 71 years today, with of course huge variations between and across countries. But what is most telling is the changing shape of populations. Anytime now, the number of people aged over 65 years worldwide is set to exceed those under the age of five for the first time in human history, and is projected to be almost twice the number of the very young by 2050. The fastest growing segment of the population is in fact the very elderly, with the proportion of people aged 85 and over projected to increase by more than 150 per cent between 2005 and 2030, compared with 104 per cent for those over 65 years and only 25 per cent for those below this age. By the mid-century, the number of people over the age of 100 is expected to be around 10 times what it was in 2010.

The big question is: what will life be like for us as we reach these venerable ages? No matter how positive and philosophical one's general disposition, one cannot ignore the evidence that for too many of us old age is nasty, brutish and long. A five-year-old girl in the UK today can expect to live to a little over 80 years. But evidence suggests that her last 19 or 20 years will likely be dogged by ill health. For a boy born in the same period, the expected lifespan is just under 80 years, but his 'health-span' is 63 years.

In a provocative essay written in 2014 explaining why he hopes to die at 75, the American oncologist Ezekiel Emanuel reviews the evidence of research and concurs with gerontologist Eileen Crimmins of the University of Southern California that, 'Over the past 50 years, health care hasn't slowed the ageing process so much as it has slowed the dying process.'

Now in my late sixties and still swimming happily in the mainstream of life, I am reminded constantly of the inexorability of ageing with the creaking of my joints as

I get out of bed in the morning or emerge from the car after hours behind the wheel. As I linger to stretch my stiff limbs, I think to myself, oh for a squirt of oil like I give my bike to keep the gears running smoothly! I see my mother as a model of what probably lies ahead for me, if I don't succumb to sudden catastrophic illness. Fit, positive, mobile and on the ball into her early nineties, I watched her lose her sight, her hearing, her beloved life partner and most of her friends, and finally her mind, across her ninth and tenth decades of life. And as I sat with her in her final years, simply keeping her occasional company along with my two sisters, I couldn't dispel the image of her once vibrant spirit like a little bird trapped in a ruined building, occasionally beating its wings in anguish against the crumbling walls from which it could find no escape. 'I've had enough; why do I simply keep on going?' she'd ask plaintively in her more aware moments. It's a question that nags us with increasing urgency as the years roll by and the prospect of prolonged decrepitude becomes harder to ignore.

But what if there was some common mechanism or mechanisms underlying the diseases of old age – mechanisms that we could tinker with to prevent or delay such crippling conditions, so that we could remain fit and active and independent till deep into old age? This is the real promise of gerontology. Yet the message is drowned out by the more bizarre claims of the 'immortalists' and 'transhumanists' – among them, it must be said, some really smart scientists – preoccupied with achieving extreme longevity and even cheating death itself. So beguiled are the media by such visions that this is where public debate about gerontology – or geroscience, to give it its latest name – tends to get stuck. The very real progress in teasing out and tinkering with the roots of age-related disease has been ignored, even as the crisis becomes obvious to everyone: the UK's National Health Service (NHS) is broken, and the debate about who should pay for the care of old people and how is rancorous.

At a conference on ageing held in New York in June 2017, Richard Faragher, Professor of Biogerontology at the University of Brighton, threw up a slide on the screen behind him. 'This is not an eyesight test,' he quipped, indicating a graph showing five diminishing columns, left to right. The tallest was marked '£715 billion: UK budget'; the next one down '£106 billion: NHS'; the next '£42 billion: spent on people 65 years and over'; and the next '£10 billion: science budget'. No one could see the final column, which was no bigger than a full stop. 'That's research into the basic biology of ageing,' he said. '£0.2 billion. Three times fuck all.' In other words, we spend next to nothing on studying the common causes of diseases that consume nearly half the NHS budget.

But my book is not going to be preoccupied with the politics of ageing, nor yet with the whackier aspects of geroscience. There are already a number of good books out there dealing with the 'could we live to 150, 500, 1,000 years or more – and would we want to?' type questions. Here I shall turn my attention inwards, for I have seen a shrunken and lacy old brain under a microscope; watched, massively magnified, elderly immune cells that have lost their satnavs (GPS), zigzagging drunkenly towards a site of injury; and seen old blood vessels like perished knicker elastic in a display cabinet at one of Gunther von Hagens's *Body Worlds* exhibitions. These are images of old age deep within our bodies, and they are transfixing. But what are the mechanisms that produce these effects? Taking my cue from Von Hagens, I shall be peeling back the skin (figuratively speaking) and separating out the sinews, muscles, bones and organs to look for answers to why and how our skin is wrinkling, our hair turning grey and our wounds taking much longer to heal than they did when we were kids; answers to why we are falling further and further to the back of the group in cycle rides and hikes, and why words escape us at crucial moments in conversation.

Scientists will always be driven by intense curiosity to explain the world around and within us, if with no other goal than to increase the sum of human knowledge. But for many geroscientists there is a more urgent sense of purpose. Ageing, says Tom Kirkwood, who has worked in this field since the early 1970s, 'is a fundamentally important process that is underpinning one of the biggest societal changes on the planet'. He and his fellows in this field have no doubt that their research holds the key to saving society from the crippling costs of care, and us, as individuals, from the fearsome and protracted indignities of advanced years.

'I feel like I spend every day rescuing people drowning in a river,' lamented a doctor on the front line of patient care in the UK's National Health Service. 'I save as many as I can, but they keep coming. It's exhausting. Eventually you just want to get out, walk upstream and stop the bastard who keeps pushing them in.' That, in a nutshell, is what geroscience is aiming to do, too, and what this book is about. But I am working on a vast canvas: each chapter could merit a book in itself, and all I can hope to do here is to sketch, in broad brushstrokes, some of the most interesting and important topics to stimulate wider curiosity in the very real prospect of healthier old age. Already in labs across the world, clocks have been turned back on failing tissues, and life has been extended – often dramatically – in many different organisms. The biology is telling us clearly that there's a great deal we can do to slow down and ameliorate the inexorable process of ageing.

CHAPTER ONE

What is ageing?

Biology is restless, never still. Our bodies are changing constantly in response to signals from within, and from the outside world. 'As a result of this unremitting change that begins at conception,' says biologist Richard Walker, 'ageing, the seed of death, is planted within each of us on the day we are given life.' In his book, *Why We Age: Insight into the Cause of Growing Old*, Walker describes growing up in America in the 1950s and '60s, an enthusiastic hippy in pursuit of youthful ideals, freedom and fun. But unlike most of his peers, he harboured a deep fear, even resentment of old age. 'One of the greatest wonders of youth,' he writes, 'is that there are really no limits to the things that your mind thinks you can achieve. So one evening while riding in my classic 1954 MG model TF with the top down and flush with the bloom of youth in body and spirit, I decided to find the cure for ageing.'

But the question then, as now, was: at what stage do the continuous changes in our bodies stop being constructive – driving tissues and organs towards maturity and optimal function and organisms towards harmony with their environment – and instead become destructive? In other words, what is ageing?

'Ageing is the universal, progressive and intrinsic accumulation of deleterious changes,' says one gerontologist. 'Ageing is the gradual failure of health maintenance systems in our bodies,' says another. 'Ageing is a disease – or a disease super-syndrome, if you like,' says another. 'I think damage over time is what ageing really is.' 'It's dying from the inside.'

Without consensus or a clear definition of when ageing begins, how it happens and why, scientists studying the process are left shooting at moving targets in the mist, trying to deduce the rules of a game that is being played out before their eyes. It is not surprising, therefore, that dealing with the ravages of age has focused on addressing the individual diseases – conditions such as cancer, heart failure and dementia – that are clearly and unequivocally pathological. Almost nowhere in medical training, let alone in the popular mind, is there appreciation of the fact that ageing itself might be the problem – that these diseases are the symptoms, the end game – and that just because ageing is a natural process that happens to us all inexorably if we avoid early death, it doesn't mean that it is either healthy or intractable.

The Greek philosopher and scientist Aristotle, alive in the fourth century BC, believed ageing was the result of gradual cooling of internal organs – in other words, the quenching of some internal flame. The ancient Chinese believed it was the result of imbalance or loss of a vital essence stored in the kidneys that sustains all bodily functions. This idea underpins traditional Chinese medicine today, which prescribes acupuncture, special foods and herbal concoctions to restore the body's balance between yin and yang – its passive and active life forces – to maintain health and youth. So, too, all kinds of present-day practices, such as yoga, meditation, massage with aromatic oils and the taking of herbal infusions, have their roots in ancient beliefs and customs from India about how to ward off time's depredations.

The first modern theory of ageing was proposed in the late nineteenth century by August Weismann, a German biologist considered by some to be one of the most important evolutionary thinkers of his time. In a nutshell, Weismann suggested that our biology could not withstand indefinitely the constant barrage of insults and injuries of daily living, and that Nature's solution was to replace worn-out bodies with new, undamaged ones. He came up with the idea that the inheritance of traits is passed on in 'immortal' germ cells (sperm and ova) and that the cells of the body, known as somatic cells, take the brunt of life's insults and have a naturally limited lifespan; once the body has matured and reproduced, it begins to decline.

Weismann originally believed ageing and death to be programmed; that evolutionary forces had selected for a death mechanism that would remove damaged individuals once they had fulfilled their primary purpose of passing on the gift of life, in order to prevent competition for space and resources with succeeding generations. 'Worn-out individuals are not only valueless to the species,' he wrote in 1889, 'but they are even harmful, for they take the place of those that are sound.' Though the theory of purposeful, programmed death will forever be associated with his name, in fact Weismann began to have doubts as he himself grew older. He modified his views, suggesting that old individuals were not the burdensome nuisance he had originally believed, that their effect on the species was neutral; ageing and death may not, after all, be programmed, but the result of progressively worn-out bodies simply running out of steam at their own pace.

Evolutionary ideas dominated the field in its early days, and continue to provide the framework for much of what goes on in geroscience today. In 1952, the British biologist Peter Medawar, who won a Nobel Prize in 1960 for his work on the immune system and transplant rejection, wrote a paper setting out his theories of why we deteriorate with age. Evolution occurs as a result of random mutations in the DNA of egg and sperm cells. Over eons of time, those mutations that confer benefit, increasing our fitness to reproduce, are the ones that survive in our species, while the ones that weaken us, increasing our chances of dying before we reach maturity, or too soon thereafter to have raised many offspring, will die out.

Genes, however, are not all expressed at the same stage in life and Medawar reasoned that it is possible for a mutation to occur that doesn't reveal its ill effects until late in life – possibly even beyond the childbearing years. The later in life a gene mutation is expressed, the weaker the ability of natural selection to eliminate it, and for this reason Medawar dubbed the post-reproductive period a 'genetic dustbin'. The harmful, late-acting mutations that have accumulated in this genetic dustbin are, suggested Medawar, the drivers of ageing. Dramatic examples of such harmful dustbin genes are the ones for Huntington's disease and familial Alzheimer's, both of which cause deadly degeneration of the brain that typically develops in later life.

Just five years after Medawar's paper, in 1957, the American evolutionary biologist George Williams came up with a deeper, more sophisticated version of this same theory. A single gene can have multiple effects in the body, depending on where and when it is expressed – a phenomenon known as pleiotropy. This multipurpose characteristic of genes helps explain why such a complex organism as ourselves can be produced by only about 20,000 genes – hardly more than the microscopic worm *Caenorhabditis elegans* (*C. elegans*) that is so popular as a model organism in biology labs.

Williams suggested that a gene mutation that has beneficial effects early in life might have harmful effects later in life, and this he termed 'antagonistic pleiotropy' – an ugly bit of biological jargon that cannot be avoided because it crops up all over the show in gerontology research. As in Medawar's 'mutation accumulation' theory, the harmful effects of the mutation would be hidden from the forces of natural selection because they don't compromise reproduction. Or, as Williams himself put it: 'natural selection will frequently maximize vigor in youth at the expense of vigor later on and thereby produce a declining vigor (ageing) during adult life.' Unsurprisingly, some have dubbed this, more graphically, the 'pay later theory'.

Williams gave two graphic examples of his idea. One involves the calcium circulating in your blood. You need this to draw on freely when you're young, to build and remodel your skeleton and to mend broken bones quickly so that you're not crippled and vulnerable. This would have been essential to survival for our hunter-gatherer ancestors. However, if you get to be 65 or 70 – as you so rarely did in pre-modern times – all that calcium in your blood begins to settle in your vascular system and you get hardening of the arteries, a classic condition of old age. But that is of no consequence to the forces of evolution: you will have had your babies already and done your bit for the species.

The other example Williams liked to give featured testosterone, the sex hormone responsible for the growth of the prostate, the gland at the base of the penis that supplies fluid to protect and nourish the sperm. Genetic variants that encourage overproduction of this hormone may also stimulate overabundant growth of the prostate that, in younger men, may increase their sex drive and reproductive success, giving them an advantage in the natural selection stakes. But it frequently causes problems for older men – most commonly, difficulty with urinating because it puts pressure on the bladder and the tubes leading out of it, and prostate cancer as the errors build up in the continuously dividing cells.

Fast-forward 20 years to the late 1970s and Tom Kirkwood, a mathematician working in medical research on blood disorders, has been thinking about one of the mysteries of cell division seen in lab dishes – namely that they inevitably grow old and die after a certain period of time. His interest has been stimulated by a chance meeting at work with a molecular biologist, Robin Holliday, who sought his help with modelling how errors in copying the DNA between one generation of cells and the next might build up. Could this be the key to ageing in us? Ageing was way outside his usual focus on blood, but Kirkwood remained fascinated. His reading around the subject in his spare time introduced him to the ideas of August Weismann, and as his thoughts crystallised, he developed a theory of ageing that built on Weismann's distinction between the immortal germ cells, sperm and eggs, and the mortal somatic cells of the body. He published his 'disposable soma theory of ageing' in 1977 in the journal *Nature*.

In a nutshell, the argument goes like this: for an organism living in the natural world with all its hazards, the most important consideration – the biological imperative, if you like – is that it survives long enough to reproduce and nurture its offspring to independence. Maintenance of the cells as they ceaselessly divide to ensure that they do so without error is energy intensive; in an environment where resources are limited or hard-won, it makes sense to invest most heavily in maintenance of the germ cells through which life is passed on, rather than in maintenance of the soma (the body), which is only required to last until it has successfully launched the next generation.

In short, making cells immortal is extremely costly biologically, and why bother with whole organisms if they are likely sooner or later to succumb to accident, disease or predation out there in the supremely indifferent world? Natural selection is concerned only with the survival of the species, not the individual. Hence, says Kirkwood, only our germ cells – the crucibles of life – are immortal, while our bodies are 'disposable'. They age gradually as a result of lack of investment in maintenance machinery.

I first met Tom Kirkwood in the 1990s when I was making a documentary on ageing for BBC radio. So, on a crisp

February morning in 2017, I set off by train from my Edinburgh home for his office in Newcastle to find out more about the disposable soma theory – how he came up with it and whether it has stood the test of time.

Kirkwood is a quiet man who holds your attention with an unblinking gaze behind wire-rimmed specs as he talks in a slow, thoughtful manner. He was born in South Africa, where his grandfather had been a low-paid worker in the gold mines east of Johannesburg and his father, who left school at age 14, was a self-made man. Kirkwood's parents met during World War II when his mother, who grew up in Rhodesia, volunteered as a nurse in a military hospital in Nairobi to which his father had been sent from the war front in Egypt suffering from malaria. Much influenced by his experiences during the war, Kirkwood senior became heavily involved in race relations issues back home in South Africa, and in resistance to the Nationalist government that came to power in 1947 and the following year introduced the policy of racial segregation known as *apartheid*. In 1955 he moved his family to England, where he was appointed the inaugural Professor of Race Relations at Oxford University.

'Oxford in the fifties was a lovely place,' says Kirkwood. 'We had a college house which had been a large, ramshackle Victorian vicarage. It was a household of six children, and it was always open. We had friends and colleagues of my father's visiting from all over the world, but he specialised in African studies so we had a lot of people coming through from Africa – people who later became the heads of the newly independent former Commonwealth states. So it was a household that was open and full of discussion and ideas.'

Though he took a degree in maths at Cambridge University, Kirkwood had always been interested in biology, an interest nurtured by long spells back in the wild, open spaces of southern Africa as he grew up. It is not surprising, then, that ageing as a topic should have appealed to him, since mathematical and biological approaches are complementary disciplines in unravelling its deep mysteries. 'I can remember very vividly how I suddenly saw the implication of the work I'd been doing for the last couple of years [with Robin Holliday],' he said with a smile of reminiscence. 'It was February 1977, a cold winter's night and I was lying in the bath and musing on this, when suddenly I realised that, *of course*, the work had shown that you could avoid the propagation of errors if you invested enough energy in error suppression.'

Kirkwood had been musing, too, on August Weismann's ideas about the distinction between egg cells and body cells, and lying back in his bath that February night he suddenly realised how the two trains of thought fit together. 'It would be worth investing [in good-quality error suppression] in the germline. Indeed you'd *have to* do it in the germline … If you hadn't evolved to do that in the germline we wouldn't be here today,' he explained. 'But then for the rest of the cells of the body, perhaps this is just too expensive. The vast majority of animals in the wild die young – very few of them make it through to the kind of age when ageing is itself a problem – so all you need is enough maintenance to keep the body in decent shape [until it has reproduced].'

This was the seed of the 'disposable soma' concept. Excited, Kirkwood got out of the bath and scribbled his idea on a piece of paper lest he forget it while on duty travel to Sweden the next day. When he came back he worked on it, then wrote it up as a scientific paper proposing a new theory. 'I was so new to science and I hadn't had the benefit of a conventional scientific training,' he explained. 'So I wanted to run it by key people who wouldn't have hesitated to tell me that I was barking mad – or that it had all been done before!' He ran the idea past Robin Holliday, past Leslie Orgel, a British chemist known for his theories on the origins of life, and past John Maynard Smith, whom Kirkwood considered 'the greatest evolutionary biologist of his day', with whom he had had some contact already.

'They all really liked the idea, so it was published in 1977, and it was quite interesting, the reaction to it,' he said. 'A couple of years later I was at my first international conference on ageing in the States. There was an American gerontologist who got a little drunk in the bar and he prodded me in the chest and said, "Tom, your paper in *Nature*, we did that with my journal club* a few months ago with my students and we haven't laughed as much in years!" So I don't think the idea immediately took off …'

Kirkwood's theory begs an obvious question: if ageing and death are consequences of built-in obsolescence – a strategy of investing only enough in maintenance of body cells to ensure a good chance of launching the next generation – do long-lived species invest more in maintenance of their bodies than short-lived species? In 1977 such questions could not be tested. But technology has advanced at startling speed, so that today researchers can watch what is going on in single cells in real time. In 1999 one of Kirkwood's PhD students, Pankaj Kapahi (whom we'll meet again later in our story), set out to test the disposable soma theory for his doctoral thesis. He took skin samples from eight different mammal species with very diverse lifespans, grew their cells in lab dishes and threw bad things at them. The prediction was that the cells from long-lived species would be able to handle the bad stuff better than the short-lived species, and that's exactly what Kapahi saw.

---

* Journal clubs are groups of people who get together regularly and usually informally to critically evaluate interesting articles in the academic literature relevant to their discipline.

'The theory was beautifully confirmed,' smiles Kirkwood. 'Kapahi's work acted as a benchmark for a whole series of subsequent studies that have tested this theory in a variety of ways, and they have confirmed time and again that there is this fundamental property – that longevity is bought, effectively, by investing in better maintenance and repair.'

In 2004, scientists working on embryonic stem cells made a very interesting discovery that further endorsed the disposable soma theory. Embryonic stem cells can be programmed to become any type of specialised cell that is required in the body. The researchers found that these earliest precursors of all other cells are, like the germ cells, immortal: they too can proliferate indefinitely. But what most excited Kirkwood and the disposable soma people was the revelation that within days of an embryonic stem cell being programmed to produce a specialised body cell (a process known as *differentiation*), the whole suite of maintenance mechanisms is downgraded. These mechanisms include the DNA repair tools, and the antioxidant defences that protect our cells from the harmful by-products of metabolism (burning sugars to produce energy). 'For me this was an absolutely wonderful moment,' says Kirkwood, 'because in the original disposable soma paper, I had predicted that the energy-saving strategy that would reduce the investment in suppression of errors should occur at or around the time of differentiation of the soma from the germline.' A pause as he looks back over time, and then he chuckles. 'You know there are very few moments in science when you can say, "I told you so!"'

Kirkwood's theory suggests an answer to another intriguing question: since all animals are made up of the same cells, the same basic building blocks, why is there such huge variety in lifespans across species? The disposable soma theory suggests that the degree of investment in maintaining the body of any creature, and thus its lifespan, is determined by its environment. If its chances of survival are short, versions of genes that promote swift maturity and reproduction will be favoured by natural selection over those that delay these vital life events. Thus mice, which are highly vulnerable to predators, typically live only a matter of months in the wild, whereas the equally tiny pipistrelle bat, which can evade predators with its aerial acrobatics, can live to around 16 years.

Although in this disputatious field of science it still has its critics, Kirkwood's 'disposable soma theory', which he has further refined over the years, provides a plausible explanation for *why* ageing occurs, and a framework for many of the ideas that have been developed since about *how* it happens. In 2013 a bunch of scientists working on diverse topics relevant to ageing decided to draw up a list of 'hallmarks of ageing' – characteristics of the elderly body that 'represent common denominators of ageing in different organisms, with special emphasis on mammalian ageing' – to help bring conceptual clarity to the field and to guide research. In so doing they were taking a leaf from the book of cancer research, which gained major momentum after a similar exercise in 2000 when two cancer scientists, frustrated by the scattergun approach of their field, drew up a list of six defining features (expanded to 10 in 2011) known as 'the hallmarks of cancer'.

In drawing up their hallmarks of ageing, the scientists, led by Carlos López-Otín of the University of Oviedo in Spain, set three essential criteria: that a characteristic should be manifest in normal ageing; that if it were aggravated under experimental conditions, it would accelerate the normal ageing process; and that if it were ameliorated under experimental conditions, the normal ageing process would be retarded and lifespan increased.

The nine characteristics that fit this bill are:

- **Instability of the genome.** This is a result of the accumulation of genetic damage throughout life, and can be caused by all manner of things intrinsic and extrinsic to the cells, such as errors in copying the DNA during division, activity of the toxic by-products of energy production within cells, or physical, chemical or biological threats from the outside.
- **Telomere attrition**. Progressive shortening of the telomeres, which are the protective caps on the ends of chromosomes often described as being like the little plastic tips on shoelaces. Every time a cell divides and its chromosomes are copied, a few bits are lopped off the ends, and the telomeres get shorter. When they get too short for the stability of the chromosome, the cell will stop dividing and its nature and function will change.
- **Epigenetic alterations**. Each cell contains a full complement of genes contained in our DNA, but the individual genes are only activated when and where they have a job to do. Otherwise they sit there in the DNA doing nothing. The action of the genes is orchestrated by chemical compounds and proteins that can attach to the DNA and switch genes on or off and modulate their activity. Together these chemical compounds and proteins comprise the 'epigenome' (meaning 'beyond the genome'), which throughout life accumulates defects that in turn affect the activity of the genes.
- **Loss of proteostasis**. Cells contain a vast abundance of proteins, which are the products of activated genes and which carry out almost all the tasks in our bodies. Proteostasis is the process by which the cell brings order to this potentially unruly mob of individual proteins, which are otherwise all focused on their own goals.
- **Deregulation of nutrient sensing**. Cells have evolved exquisite mechanisms for adjusting their

behaviour to make the most of the nutrients available for generating energy and providing raw materials for growth. These mechanisms rely on sensors that constantly relay signals about the body's nutrients status.

- **Mitochondrial dysfunction**. The mitochondria are the cells' batteries. They are organelles found in large numbers in all mammalian cells except the mature red blood cells, and their main task is to take in nutrients (sugars and fats) from the cell and break them down to produce energy.
- **Cellular senescence**. Cells that normally divide lose their ability to do so after a certain number of divisions, as measured by shortening telomeres on the ends of their chromosomes. They then enter a state of permanent arrest known as senescence. Besides shortened telomeres, other forces, such as irreparable damage to the DNA or epigenetic alterations, can also cause cells to senesce.
- **Stem-cell exhaustion**. Adult stem cells are undifferentiated cells kept in reserve for repair and maintenance of the body. They are found tucked away in most tissues and organs, and can be programmed to replace lost or damaged cells in the tissue in which they are found. Over the years, these reserves get run down.
- **Altered communication between the cells of the body**. This is a result primarily of chronic, low-grade inflammation of the tissues.

These hallmarks describe the common, universal characteristics of ageing – and they provide strong reference points for the various researchers as they roll up their sleeves and get on with the investigation. But what researchers following any of these paths share with their fellow researchers travelling different paths is the desire to know what kicks the whole thing off, or what and where are the master switches of the ageing process.

Musing about the origins of organic life, which had been his obsession, the brilliant British chemist Leslie Orgel called it 'chaotic intellectual territory'. Much the same could be said about ageing. But a combination of burning and sometimes brilliant brains and rapidly evolving technology is affording us some fascinating and important insights into what is happening deep within our bodies and beginning to make sense of the great mysteries of ageing and death.

CHAPTER TWO

Wear and tear?

The idea that our bodies wear out, and succumb to the forces of entropy like everything else we see around us – our cars, our houses, our furniture, our clothes and electrical devices, as well as our dogs, cats and budgies, and the flowers and trees in our gardens – makes intuitive sense to most of us not trained to investigate such matters. It has dominated the field of gerontology in one form or another since August Weismann developed his theory of why we age in the 1880s. But how might it happen?

In 1954 the American biochemist Denham Harman was asking the same question, and he came up with the 'free radical theory of ageing' (this is also known, somewhat confusingly, as 'the oxidative damage theory'). This proposed that free radicals – the by-products of chemical processes within our bodies, including metabolism which converts our food into energy using oxygen – are toxic and can cause havoc in our cells. We have good defences against free radicals, the vast majority of which are deactivated or mopped up by specialised scavenger cells, while damaged cells are killed and recycled. But as the processes of energy production and waste management decline in efficiency over time, free radicals proliferate and do increasing damage.

Harman, born in 1916 in San Francisco, trained as a chemist and worked for some years as a research scientist for the Shell Oil Company. But he had a deep curiosity about life, and at the age of 33 years he went back to school to study medicine. He was especially curious about why everything dies, and the questions raised by the atomic bombing of Hiroshima and Nagasaki by the US in August 1945 gave him a clue. At the time of the bombing, very little was known about the effects of non-lethal doses of radiation on the human body, and so, when World War II ended, America and Japan signed a joint agreement to study the effects on survivors of the atomic bombs that had killed an estimated 130,000–230,000 Japanese citizens within the first few months of the attack by US warplanes. America was particularly anxious to find ways of protecting soldiers and civilians from the effects of any future conflict that might involve nuclear weapons.

The investigators found that mice given high doses of radiation produced enormous quantities of free radicals that overwhelmed their normal defences and were responsible for the toxic effects of radiation. Intriguingly, too, these toxic particles seemed to age the mice prematurely. Harman was familiar with the action of free radicals in non-organic materials from his days in the oil industry. As he followed the investigation into their effects in living creatures he became convinced that the free radicals generated by our own normal biological processes are the cause of ageing. It was a revolutionary idea: free radicals were generally believed to be too toxic to exist naturally in living things.

So what exactly are they? Free radicals are atoms that have, in the course of the chemical reactions within cells that sustain life, lost electrons and become highly unstable. These delinquent atoms careen around inside the cells until they are able to restore their electromagnetic balance by ripping electrons from other atoms, often starting a chain reaction. Free radicals 'burn like gunpowder, until hundreds of thousands [of atoms] are damaged', Mikhail Shchepinov told *New Scientist*. They play havoc with the membranes and contents of cells. And because of their unbalanced electrical charge, they're also drawn like magnets to DNA, where they stick to the ribbon of genetic material and cause random mutations.

This effect on DNA is a double-edged sword. It disrupts the activity of genes and can cause cancer and other diseases. But it also makes free radicals key agents of evolution, since it is through natural selection working on genetic mutations that we are able to adapt to changing environments. Free radicals assist in the crosstalk between cells; they can, under certain circumstances, prime cells to be less vulnerable to stress, and they might even have a role in fighting bacteria and viruses. But on the whole they are bad news, and our bodies put up strong defences against them. Scavenger cells of the immune system mop up almost all of them, and it is the gradual build-up of damage by those that are not cleared in time that Harman proposed as the cause of ageing.

He supported his hypothesis by demonstrating that he could increase the lifespan of laboratory mice by up to 30 per cent by giving them drugs to protect them against radiation. He could also increase their lifespan, but not so dramatically, by giving them antioxidants designed to prevent oxidative damage. The relative weakness of antioxidant therapy perplexed Harman for a long time and led him eventually to conclude that most free radicals are generated inside the mitochondria, the cell's batteries, which burn calories to produce our energy and which are impenetrable to compounds introduced from outside. In the 1970s he modified his theory to suggest that the mitochondria might be the body's clock, determining how long we live according to how hard we work our batteries and how much wear and tear they sustain.

But his ideas were slow to catch on, and Harman was frustrated about the fatalism that deadened curiosity – among scientists and the general public alike – about ageing as a biological phenomenon worth studying. In 1970 he founded the American Aging Association to kick-start serious research in the field, and in 1985 he helped set up the International Association of Biomedical Gerontology. As the scientific community began to wake up to the possibilities of ageing research, and as evidence for Harman's ideas accumulated with the development of ever more sophisticated technology for biological exploration, the free radical theory of ageing took centre stage, strongly influencing the direction of research until well into the twenty-first century.

Harman took to heart what he learnt in the lab about healthy ageing: he never smoked, was a moderate drinker, watched his weight and took lots of exercise, running 3.2km (2 miles) a day till he was 82, and only then slowing his pace to a walk after a back injury put paid to his running days. He died in 2014, aged 98, but lived long enough to see his free radical (or oxidative damage) theory of ageing knocked from its pedestal.

'When I came into the field 20 years ago, my impression from my colleagues was that the oxidative damage theory was pretty much a done deal,' said geneticist David Gems when I visited him in his office at University College London where he is Professor of Biogerontology. '[The attitude was that] "so many papers have been published by now that we're all agreed this must be correct". But I suspect it's a folk theory.'

Gems is renowned for pushing at intellectual boundaries as much as he is for his colourful life story (friends tell of a one-time punk, who worked in an Icelandic fish-packing plant, spent time with the Sandinistas in Nicaragua, dug graves in Guatemala and bummed around Soviet Russia in the 1980s). The oxidative damage theory, Gems believes, has persisted because of its intuitive appeal – like the idea, so long held, that the sun goes round the Earth rather than vice versa, because otherwise we would be tumbling around with all Earth's contents as our planet hurtled through space. 'It took until the fifteenth century to work that out,' he comments. 'But that's how science works. You start off with intuitions based on common sense, and it's only when you do experiments that you discover that those things that seem so intuitive are actually incorrect.'

Gems doesn't dismiss the oxidative damage theory outright, but says that since the early 2000s labs around the world, including his own, have been 'testing to destruction' the theory and its predictions and found them wanting. In the search for truth, scientists have had to wade through a veritable blizzard of data generated by studies in yeast, microscopic worms, fruit flies and mice – the traditional workhorses among model organisms of biological research – which have had their antioxidant defences knocked out or increased by drugs or genetic engineering. 'The critical issue,' says Gems, 'is that if you manipulate the levels of oxidative damage, you should see effects on ageing and on lifespan. And many studies were reported that didn't support that theory. Including human studies… experiments with people taking antioxidant supplements and looking at mortality rates, and it didn't make a difference. In some cases taking antioxidants actually *increased* mortality rates slightly.'

Many of the data were buried in obscure journals read and understood only by geeks. But one particular piece of research caused a media frenzy when what looked like a dramatic effect of an antioxidant drug given to laboratory worms was published in the prestigious journal *Science* by Gordon Lithgow and Simon Melov, then working at Manchester. The two scientists have since moved to California, and when I visited Lithgow at the Buck Institute for Research on Aging – a modern building, all space and light, which sits in wooded grounds on a hilltop near San Francisco – he pulled a fat folder of press clippings from his shelf as he told me the story of the drug that had dramatically extended the lifespan of his worms.

'The BBC came and made a documentary; Channel 4 made a documentary ... It went on for three or four years, this constant media interest in these scientists who had extended lifespan with a drug,' he commented. 'It was an antioxidant, so it was converting those free radicals into something neutral, and what we thought was going on was detoxification – the worms were very resistant to oxidative stress.' Lithgow and his team could squirt the worms with the highly toxic weedkiller Paraquat – an oxidant notorious for being the most common agent in suicides, especially among destitute farmers, in Asia – and the worms wouldn't flinch. No wonder the media were excited: it looked like the scientists had found a key to ageing, if not *the* key, and it was a surprisingly simple mechanism.

Their fellow scientists were intrigued too, says Lithgow. 'My friend David Gems at University College London said, "Can we have a go with these compounds, because we want to test out a hypothesis?" And we said, "Course you can; let's do it!" A few months go by, and he calls and says, "I can't get them to work! These are not extending lifespan in our hands."' Lithgow was alarmed and frustrated. But despite heroic efforts and much agonised discussion between the labs over the next couple of years about exactly how each group was doing its experiments, Gems and others who also tested Lithgow and Melov's drug could not get it to extend the lives of their worms. Gems even tried grinding his worms up to prove they had taken the drug on board and to measure the antioxidant activity in the extract, which he found was indeed higher than normal. But it made no difference to his lifespan results and he eventually published his findings.

The conflicting results left the researchers scratching their heads in perplexity, but it comes down to the fact that 'ageing is messy. Ageing is *really* messy biology,' says Lithgow. There are few straight lines between cause and effect in living things because there tend to be back-up systems, or else biology tinkers around to compensate for things that fail. Today Lithgow's lab is part of a project set up to standardise research using worms, and to run their tests with promising chemicals in three widely dispersed sites to ensure the results are as robust as possible before publishing.

'It took us a year and a half to standardise the protocols,' he commented. 'We'd have these agonising teleconferences, where we'd talk about how we'd pick the worms up and put them down on the plate ... We went into it assuming that almost any difference would be important. We thought, let's all buy the same make and model of incubators; let's buy the agar for the agar plates in bulk, and grow up one big batch of worms and distribute them between the labs. So everything that was humanly possible was standardised.'

In the process of controlling as many variables as they could think of, Lithgow and his colleagues recently discovered a quirk about worm biology they could never have anticipated – and that may well be the clue to why Gems and others could not reproduce the Manchester group's results with the life-extending drug. They observed that certain wild strains of worms described as long-lived on one occasion when they were being monitored would, on another occasion, be categorised as short-lived, though they were from the same population, with the same genetic heritage and raised in identical conditions. A while later, a cohort from the same population of worms would be back to their long-lived selves. The baffled scientists asked themselves: could these weird findings be down to the phase of the moon or the time of day they were being monitored? Or something to do with the technicians handling the critters? But no, all three labs observed the same clear phenomenon – a propensity to be long-lived at one point and short-lived at another, but, significantly, nothing in between. 'There's something going on in all three labs that is changing, something about the worms' metabolism perhaps – we have no idea,' said Lithgow. 'It's like dark matter! You know that dark matter is there because it affects other matter, but ...'

In fact, failure to reproduce dramatic research results is much more common than you'd think from the media coverage of so-called breakthroughs (journalists often don't follow up on a story when it gets messy or the doubts creep in), and this case might hold another lesson: that failure doesn't necessarily imply bad science or errors by either party; rather that understanding of what's going on at a fundamental level is still pretty limited. It's a reminder too that the search for universal rules in biology is fraught with pitfalls and false promises.

The oxidative damage/free radical theory was dealt another serious blow in 2009 by researchers working at the Oklahoma Medical Center. Arlan Richardson and Holly van Remmen tinkered with the genes of their laboratory mice so that they produced an overabundance of antioxidants that did a very good job of mopping up free radicals – but without any effect on the lifespan of the mice. They also did the reverse experiment, working with mice that had the genes for two of the most important antioxidants deleted from their DNA. Sure enough, the mice sustained extensive damage to their cells from free radicals, but without this shortening their lifespans – at least not in the mice that avoided developing cancer as a result of the damage.

After that, Arlan Richardson was wont to turn up at gerontology conferences and put a big red X through the oxidative stress hypothesis, remembers Lithgow. 'He was doing it to be provocative, but the entire field just went, "Oh, okay, I guess it's not that then. Good, well we can study other things."' For a long time the gerontology community had been reluctant to let go of a cornerstone of their understanding of ageing, even as the evidence grew. But then the consensus 'suddenly popped like a bubble,' says David Gems. 'It's very strange ... Fifteen years ago you'd go to ageing meetings and you'd have talk after talk referring everything to oxidative damage. Now you can go to conferences and you hear almost nothing about it.'

This does not mean that wear and tear are no longer seen as key features of ageing – they clearly are. And Gordon Lithgow remains intrigued by his original experiment with the worms, where he managed to extend their lifespan considerably by making them resistant to oxidative stress – an experiment he has repeated under much more stringently controlled conditions and with the same positive result. But what is a cause and what is a consequence of ageing? And how much of a role does oxidative damage play? Few would argue today that the by-products of the cellular processes that keep us alive are what drive us inexorably towards the grave.

In a review paper from 2009 that evokes an image of buzzards circling to land and pick at the carcass of the oxidative damage theory, Gems and his fellow scientist Ryan Doonan conclude that its decline 'represents an exciting departure, marking a new beginning for biogerontology. It is time to start thinking about ageing in new ways.'

'The crisis of the oxidative damage theory was a big moment for me,' commented David Gems later. 'It was as if your mind was freed from the necessity of thinking always in terms of damage maintenance.' So where are Gems and others looking now for explanations about how we age?

CHAPTER THREE

Telomeres – measuring the lifetime of cells

The power of a paradigm to dominate thinking and blind adherents to other possibilities has never been better demonstrated, in the ageing field at least, than by the story of Leonard Hayflick and his discovery of the limited lifespan of our cells.

Hayflick, who studied microbiology at the University of Pennsylvania, was just turning 30 when he went to work at the Wistar Institute in Philadelphia as a cell culturist – a skill he had learnt under the tutelage of one of the country's pre-eminent practitioners of the art, Charles Pomerat. The main focus of the Wistar at that time was virus research, especially for the purposes of developing vaccines against polio and other such scourges. Viruses are the ultimate parasites, unable to live outside of other living cells, whose machinery they hijack to make copies of themselves. Hayflick's job was to keep up a good supply of cells on which to grow these microbes.

For a long time, the scientists working to develop a polio vaccine had been using cells from monkey kidneys to culture their viruses. But this was not ideal: their cultures were at risk of contamination from all manner of hidden nasties in the monkey cells. So Hayflick, looking for a safer alternative, set about trying to culture human cells – and in particular, cells from human foetuses that had not been exposed to the outside world and were likely to be pristine. He sourced his supply of foetal material from a former colleague now working in Stockholm, Sweden, where they are less squeamish than in the US about abortion and the idea of harvesting tissue from foetuses. The tissue – mainly from the kidneys and lungs – would arrive by air, packed in ice, in the regular post. Hayflick would shave off wafer-thin flakes, which he soaked in enzymes that digested the connective tissue holding the cells together, leaving him with pure cells to grow on his agar gels, fed with his own magic formula of nutrients. As the cells divided in the body warmth of the incubators, eventually to cover the surface of the gel in their glass flasks, his technicians would divide them up into new flasks, and harvest them to supply the virologists. Sealed ampoules of cells were also put in the deep freeze to suspend their animation for later use or distribution to other labs.

Culturing the cells from such temperamental tissue was a grindingly painstaking process. Hayflick had a flair for it, and over time he found his cell lines in great demand from scientists everywhere. But as his intimate knowledge of his cell cultures grew, Hayflick became aware that, at some point, the cells would stop dividing and that this seemed to be a predictable event that occurred after about 50 population doublings. He was particularly intrigued by the fact that the cells didn't die; they continued to metabolise and could live in that non-dividing state for a year or more.

Cultured cells ceasing to grow and divide were familiar to research scientists everywhere. But the dogma at the time was that such cells had the capacity to divide indefinitely, and that failure to do so pointed to technical error – to contamination, inadequate nutrition or poor handling by technicians. This notion had been promulgated by Alexis Carrel, a French-born surgeon at the Rockefeller Institute in New York who was awarded a Nobel Prize in 1912 for pioneering a technique for stitching severed blood vessels. Carrel was supposed to have kept tissue from a chick embryo heart alive in a flask for well over 20 years by feeding it a regular supply of nutrients. No one else could come anywhere near repeating his experiment and all kinds of explanations have been put forward for the extraordinary longevity of his culture – including the idea that his technicians, fearful of being held responsible for killing off the eminent surgeon's precious cells that he had declared everlasting, had been replacing them secretly whenever they died. And inadvertently building a powerful myth as they did so!

But Hayflick's personal observations of cells ceasing to divide after a certain number of divisions sowed serious doubts in his mind: this looked to him like an intrinsic characteristic of the cells – even a natural process of ageing – not something environmental, and he enlisted the help of cytogeneticist* Paul Moorhead to test his theory. The two scientists needed to rule out the possibility that the changed state of the cells was due to contamination with microbes, or some unknown feature of the culture medium. They were able to distinguish between female and male cells by their chromosomes. So they put a number of female cells that had completed 10 cycles of division into a flask with male cells that had already been through 40 cycles. As Hayflick anticipated, he and Moorhead found that after another 20 cycles, only the female cells were still going strong; the older, male, cells were in the strange 'quiescent' state. All had been living under the same conditions, so it was clearly not something to do with their environment, and Hayflick and Moorhead submitted their findings to the *Journal of Experimental Medicine* – 'the Cadillac of journals' for cell biologists, according to Hayflick.

The journal's editor, virologist Peyton Rous, was himself very ready to shoot at sacred cows in the field of cancer

* Cytogenetics is the study of the structure, location and function of chromosomes in cells.

research – and had won a Nobel Prize for doing so with his discovery of tumour-inducing viruses, originally among chickens. But Hayflick's theory was a challenge too far for Rous. 'The inference that death of cells is due to "senescence at the cellular level" is notably rash,' he wrote in his rather haughty letter of rejection. 'The largest fact to have come out from tissue culture in the last fifty years is that cells inherently capable of multiplying will do so indefinitely if supplied with the right milieu in vitro.' Hayflick and Moorhead's paper was eventually published, in December 1961, in the more modest journal *Experimental Cell Research*. (It's worth noting that the paper was accepted by the journal without correction – a rare occurrence in scientific publishing, and a testament to its quality despite Rous's rebuff).

To persuade the many sceptics, Hayflick offered to supply them with flasks of cells along with predictions, from his own calculations, of when they would stop dividing. His ability to forecast these events accurately opened many researchers' eyes to what had been under their noses all the time, but which they had mistakenly put down to technical error. What he had correctly identified as the natural lifespan of cells is known today as the Hayflick limit, and it's become one of the hottest areas of research in gerontology.

For Hayflick himself, however, it has been an extremely bumpy ride. In his book *Merchants of Immortality*, author Stephen S. Hall describes how the scientist became embroiled in a bitter dispute with the Wistar Institute over ownership and intellectual property rights to his most important cell line, known as WI-38. When in 1968 he left the Wistar for a posting at Stanford University, California, the headstrong Hayflick, on a whim, removed hundreds of ampoules of 'his' cells from the lab, stored in a huge grey canister of liquid nitrogen which he propped up between his kids in the back seat of his car for the long drive west across the United States.

Though he managed to set up a busy lab at Stanford with his cells, the consequences of his heist dogged his career and family life for decades and caused extreme stress. The dispute ended in 1981 when Hayflick accepted an out-of-court settlement from the National Institutes of Health (NIH), whom he had sued for raiding his Stanford lab and confiscating his cell stocks, after accusing him of stealing government property and selling it for personal gain. The federal government finally conceded Hayflick's ownership of WI-38 and his right to keep his earnings from the sale of cells. Today he still has that precious canister of WI-38 cells, which he keeps, bizarrely, in his garage at home. Alongside it is a canister containing cells derived from the amniotic fluid that had bathed his daughter, Susan, in her mother's womb, as well as prostate cancer cells from the tumour that eventually killed his one-time boss and mentor, Charles Pomerat.

Stephen Hall was taken to see the precious cache by Hayflick, whom he portrays in his book as a courteous, warm, but complex and somewhat hot-headed man whose 'most prominent physical feature, however, may be the permanent chip on his shoulder, almost a bone spur of bitterness and resentment, about the way he has been treated during his scientific career.' Another of his colleagues has described him affectionately as 'a lovable old grump and so tough they'll turn him into Bovril when he dies.'

Hayflick's game-changing discovery in 1961 raised an obvious question: how do cells know when they have reached their limit? How are they measuring out their days? Way back in the 1930s, the American botanist and cytogeneticist Barbara McClintock noticed that chromosomes without their end parts became 'sticky', and tended to attach themselves to each other or to break down.

She hypothesised that chromosomes normally have protective tips to keep them neat and separate. These hypothetical tips were given the name 'telomeres', derived from the Greek 'telos' (end) and 'meros' (part), by fellow scientist Hermann Muller, who had observed the same problem with apparently truncated chromosomes having a tendency to stick together. But it was another couple of decades before Francis Crick and James Watson – poring over the tantalising images generated by Rosalind Franklin's X-ray crystallography – teased out the structure of DNA and paved the way for detailed exploration of chromosomes. And another 25 years were to pass before the nature of telomeres – McClintock's hypothetical tips – was discovered by Elizabeth Blackburn, then at Yale University in New Haven, Connecticut.

Born in Hobart, Tasmania, Blackburn was one of seven children who came from a family of medics and scientists stretching back several generations on both sides. The family home and garden, first in the tiny town of Snug on the coast near Hobart and then Launceston in northern Tasmania, teemed with pets of all sizes, ranging from goldfish, and tadpoles in increasingly smelly jars, to budgies, hens, cats and dogs. And as a little girl she would collect ants and jellyfish while out playing.

'Perhaps arising from a fascination with animals, biology seemed the most interesting of sciences to me as a child,' Blackburn wrote in her autobiographical sketch on receiving the Nobel Prize for Medicine in 2009 for her work on telomeres. 'I was captivated by both the visual impact of science through science books written for young people, and an idea of the romance and nobility of the scientific quest.' Her early experiences in the world of science can only have reinforced such sentiments. After taking a degree in biochemistry at the University of Melbourne, she stayed on to do a masters, and writes of her professor, 'Frank Hird taught his laboratory group members the joy and aesthetics of research. He said he thought each experiment should have the beauty and simplicity of a Mozart sonata. His laboratory group, dominated by his strong personality, was cohesive and we would sometimes drive to the hilly areas outside Melbourne, all piled into his car, Mozart playing loudly on the car radio, to have an outdoor lunch picnic among trees and wildflowers.'

Blackburn did a PhD in the similarly relaxed and informal lab of British biochemist Fred Sanger at Cambridge University, in the UK. Sanger had pioneered a method of sequencing DNA that revolutionised the technology and earned him his second Nobel Prize (he is only one of four people ever to be awarded a second Nobel). His technique was used in the Human Genome Project to decipher the instruction manual for our species.

It was using the skills learnt in Sanger's lab that Blackburn, then at Yale and working with a tiny single-celled organism common in freshwater ponds, discovered the nature of telomeres. 'Studying organisms at a molecular level was totally compelling because it was moving from being a naturalist, which was the 19th-century kind of science, to being very focused and really getting to the heart of these molecules,' she said in an interview with *Discover* magazine. 'We knew they carried genetic material and that the ends of chromosomes were protected in special ways. But what did that mean? You have no clue. It was like you were trying to look at something from 400,000 miles [644,000km] up. You could see a speck on Earth, but you had no idea that if you homed in on it, it was a cat.' Investigating telomeres was especially exciting because, 'Molecularly speaking, this was uncharted territory.'

Blackburn found that these entities, the telomeres, consist of short sequences of DNA repeated a number of times – typically 20 to 70 times in her minuscule pond-water creatures – compacted and packaged in a protein cover. What's more, she noticed that the telomeres were not copied along with the rest of the DNA during cell division, but seemed to be added to the chromosomes afterwards. Blackburn had come a very long way since McClintock first suggested these caps existed, and she published her findings in 1978. But *how* the telomeres functioned remained a mystery. This she worked out in collaboration with Jack Szostak of Harvard Medical School, who was asking related questions of his model organism, yeast, and who pricked up his ears when Blackburn spoke of her findings with the pond-water creatures at a scientific conference in 1980. Mixing and stirring the genomes of their experimental models in lab dishes and watching what happened over many cell divisions, the two scientists discovered that when the DNA is copied during division, the telomeres effectively act as buffers at the end of the line for the copying machinery that runs along the ribbons of DNA. With every division, a bit of the telomere gets knocked off. And when this cap is too short to protect the chromosome from damage, the cell stops dividing and becomes senescent.

But that left the intriguing question of how, if they aren't copied along with the rest of the DNA, the telomeres are made. This bit of the jigsaw fell into place a number of years later, when Blackburn had moved to a lab of her own at the University of California, Berkeley. When you're working in science, she said in a YouTube talk for *iBiology* magazine, there comes a time when you're trying to interpret what you have observed in the light of well-established principles, 'but it just won't fit. You just can't push it into that box any more, and you have to say, well, let's entertain some other possibilities.'

One of Blackburn's graduate students, Carol Greider, had chosen to work for her doctoral thesis on the mystery of how the tiny scraps of repeating DNA are made. After travelling down a goodly number of the cul-de-sacs so familiar to research scientists, she and Blackburn decided to test the hunch that there must be some unknown substance in cells whose specific job is to build these protective caps on the chromosomes. Greider put fragments of DNA from her pond-water creatures into a test tube along with a supply of loose, random building blocks of telomeres. When she started her investigation all the DNA fragments were the same length, and she had radioactively labelled the telomere building blocks so that she could see what happened to them in her mixture.

The year was 1984, and Greider's curiosity and commitment to her research must have been intense, for she went in to check on her experiments on Christmas Day. She found that the building blocks had been sorted out into their proper sequences to make telomeres, and had been attached to the DNA fragments, which were now various different lengths. This was the first evidence that there was indeed an unknown substance at work, and the two scientists gave the substance the name of telomerase. In time it became clear that Blackburn, Greider and Szostak had discovered, in their ultra-simple model organisms, some fundamental mechanisms that operate in a huge variety of Earth's creatures, including us, and the three shared the Nobel Prize for Physiology or Medicine in 2009.

Over the years scientists have learnt that the length of our telomeres is determined to a certain extent by our genes and that it varies from one tissue and organ to another in our bodies, as the rate of cell division also varies between tissues. Our gut cells, for example, divide rapidly and are replaced roughly every four days from the pool of stem cells responsible for tissue maintenance, while the turnover time for liver cells is six months to a year. Scientists have learnt, too, that telomerase, which builds and maintains the telomeres, is only active throughout life in our 'immortal' germ cells (sperm and eggs), in stem cells and in the bone marrow supplying blood cells, but it is switched down or off in most other body cells. Telomerase can, however, be one of the drivers of cancer, when mutant genes manage to switch it on again, helping the cells to avoid senescence, no matter what the state of their DNA, and continue their rampant growth.

So could telomeres, the ticking clocks that mete out the lifespan of our cells, hold the key to how we age? Certainly, shortened telomeres and telomerase activity are involved in a number of the major age-related diseases besides cancer, including heart and lung diseases, diabetes and dementia. And there is a condition called dyskeratosis congenita which is caused by mutations in the telomerase genes, and which is sometimes classified as a progeria – or premature ageing – syndrome because it renders carriers vulnerable at very young ages to bone-marrow failure, lung problems, osteoporosis, deafness, greying and loss of hair, and tooth decay, besides the more common, milder symptoms of abnormal skin pigmentation and finger and toenail growth.

Telomere attrition is one of the hallmarks of ageing mentioned in Chapter 1, and their length can be an indicator of someone's age in certain circumstances. In the late 1990s and early 2000s, the idea that tinkering around with telomerase could be used to treat cancer, as well as restore vigour to tired cells and extend lives, got a lot of hype in the media. 'The public's fascination with the telomere model of ageing is understandable, because [it] provides a timer for cell division and is simple to comprehend,' writes gerontologist Lenny Guarente in his book *Ageless Quest*. 'But in my view, cultured cells constitute a rather unnatural system,' he cautions. 'In the whole animal, those cells would be sloughed off before their telomeres were too short and they would be replenished by differentiation of precursor cells that *do* have telomerase. The cultured cells are cut off from this source of renewal and left hanging in the breeze.'

Despite the scepticism of some in the field, the idea of telomerase therapy looked so promising to maverick US gerontologist Michael West that he set up a pharmaceutical company, Geron, to explore the possibilities. West, a pioneer in stem-cell research, tells a lovely story in *The Translational Scientist* magazine about how he invited Leonard Hayflick to visit the company. The discoverer of the natural lifespan of cells was highly sceptical about Geron's ideas, but he was persuaded to donate a patch of skin from his leg to be used for an experiment. By inserting the telomerase gene into cells extracted from the skin sample, West succeeded in making them overrun the Hayflick limit and carry on dividing. 'Hayflick was dogmatic about the fact that we would never be able to intervene in human ageing,' says West, 'but his cells ended up being the first to be immortalized.'

It was a terrific breakthrough with a nice touch of poetry. But despite this and other successful experiments with animal models such as mice, the results of telomerase therapy have been inconsistent, unpredictable and not always easy to interpret. This has driven home the message that the picture is a lot more complicated than it seemed early on, and today studying cellular senescence beyond the telomeres is providing rich new insights into the biology of ageing.

CHAPTER FOUR

Cell senescence – down but not out

When Judy Campisi, a leader in the field of cell senescence, gave her first talk on the topic at a gerontology conference in the late 1980s, she received a scathing response from 'an old curmudgeon' at the University of California, Berkeley. 'Young lady,' he said haughtily, 'you are studying a tissue culture artefact. This has nothing to do with anything *in vivo*.'

'I could have killed him for the "young lady" alone!' says Campisi with a laugh. But the old curmudgeon's attitude was not unusual. 'In the late eighties and early nineties a lot of people were just very sceptical that this process of senescence was anything other than a cell culture artefact.' For a very long time, the topic led to acrimonious exchanges at gerontology meetings. The hostility began to subside only as evidence mounted that clearing out senescent cells in a whole range of model organisms extended their lifespan by up to 30 per cent and alleviated age-related diseases.

In fact it took Campisi herself a while to see the light. A petite, soft-spoken woman with the graceful deportment of a ballet dancer, she had begun her career studying cancer – particularly 'what makes a cancer cell proliferate when it shouldn't.' Cellular senescence, the phenomenon Hayflick had recognised, was interesting to the cancer community as a possible mechanism for suppressing tumours – Nature's way of neutralising cells that might, with time, have gathered dangerous mutations in the rough and tumble of living. Hayflick had suggested such an idea when he noted that, unlike his normal foetal cells, cancer cells didn't seem to 'hit a wall' of division.

At one point in her career, Campisi was asked by two colleagues studying ageing if she would add her name and research topic to a funding application they were making because they desperately needed a third person. 'It's probably all bullshit; cell senescence has nothing to do with ageing, but you can just *say* it has!' they advised her. Campisi was happy to go along with them: anything that promised funding for her research sounded good to her.

In his various communications about his cell cultures, Hayflick had mentioned that those that had reached their replicative limit looked 'old', and he mused as to whether what he was seeing was recapitulation in a dish of what happens to us as we age. It was highly subjective speculation – after all, what exactly does 'old' look like? – and only a handful of scientists, considered nutters by their peers, took notice. But as she worked with her cells, Campisi became intrigued by the evidence that what she was witnessing was indeed a process of ageing. What's more, this was not a separate phenomenon, but was intimately connected to her main focus of enquiry, tumour suppression. Cancer and ageing, she observed, seem to be two sides of the same coin: ageing is the price we pay for protection against cancer, in that Nature's way of limiting the proliferation of potentially dangerous cells is to put a brake on their ability to divide after a certain length of time.

Here's where the story gets a little more complicated. The system protecting us from cancer does not rely on telomeres alone to limit the life of potentially dangerous cells. Many other things – including UV radiation from sunlight, oxidative stress (those free radicals again), chemicals in the environment, dangerously low levels of oxygen or nutrients, aberrant proteins clogging up the machinery – can damage our DNA and put the brakes on cell division. They do so by setting off a general alarm system controlled by genes known as 'tumour suppressors'. The most notable of these tumour suppressors is a gene called simply p53, which is present in all our cells. p53 is the recipe for a protein whose main task is to scan our cells constantly to ensure they grow and divide without making serious mistakes. If it receives signals that the DNA in a dividing cell is damaged, p53 will stop the cell in its tracks and send in the repair team before allowing it to proceed. If the damage is beyond repair, the tumour suppressor will instruct the cell to commit suicide – a process known as apoptosis – or else it will induce a state of permanent arrest, or senescence. Because of its vital role in ensuring the integrity of our DNA, p53 has been nicknamed 'the guardian of the genome' by David Lane of Dundee University, Scotland, one of the four scientists who discovered the tumour suppressor in 1979*. If this gene is working properly, it is virtually impossible to get cancer.

Senescent cells have proved to be extraordinarily stable, in lab dishes at least, says Campisi. 'Rumour has it that one lab in Texas has kept senescent cells perfectly happy in a culture dish for years – until the technician got fed up and refused to feed them!' No one yet knows how stable they are in living beings, nor how long they last. Scientists do know, however, that senescent cells are found in people of every age from babyhood upwards, and that they are normally recognised and cleared away regularly by the immune system. But as we get older, so too our immune systems age and get less efficient at clearing senescent cells just as we are producing more of the things. Thus senescent

* The other three were Arnie Levine at Princeton University, New Jersey, US; Pierre May at the Integrated Cancer Research Institute, Villejuif, France; and Lloyd Old at Memorial Sloan Kettering Cancer Center, New York, US.

cells accumulate in our tissues as the years advance, clustering particularly at sites of age-related disease such as skin ulcers, arthritic joints and furred-up arteries.

'What happens is this,' says Richard Faragher, whom we met in the introduction and who studies senescent cells at the University of Brighton. 'If you're one of those 18-year-old students I teach, the moment you make a senescent cell, it's clobbered. By the time you're my age, it's more like, "you're through to the senescent cell helpline, your call is important to us, an immune system representative will be with you shortly ... In the meantime, here is some music, don't go away!"'

So how do these cells actively promote ageing? Senescent cells are not dead but dysfunctional. As they continue to metabolise, they secrete substances into their environment that chew up collagen, the stuff that holds our cells together. Collagen fibres are long and elastic and keep our skin firm and young; as it degrades it leaves wrinkles and saggy bits. The degraded collagen also leaves spaces that give precancerous cells that may have been lurking there, held in check by firm young tissue, room to proliferate. The number of cells in tissues is fairly constant, and another theory is that the presence of large numbers of senescent cells clogging up the available space might inhibit repair and regeneration by the stem cells that are kept in reserve for maintenance purposes. In other words, they act like dogs in the manger, quite simply keeping the new guys out. The effect will be the same – poor maintenance – if it is the stem cells themselves that have sustained DNA damage and senesced.

Stem cells occupy little pockets, or 'niches', in the various tissues where they wait to be called on for repair purposes. The environment of the niche is important in keeping them fresh and healthy, and there is evidence that if any of them senesce they can alter the environment and impair the function of other stem cells in the niche. Senescent cells in the walls of blood vessels tend to forget that they're blood-vessel cells and become bone cells, leaving calcium deposits that cause hardening of the blood vessels and heightened risk of strokes and heart attacks.

But one of the most important ways in which senescent cells are thought to promote ageing is through chronic inflammation – a phenomenon so widespread and central to the process that it's become known as 'inflammaging' (pronounced 'inflamm-ageing'). The substances secreted by these dysfunctional cells include a large number of small molecules called inflammatory cytokines. 'These are normal proteins produced when there's an infection or an injury,' explained Campisi. 'The main job of inflammatory cytokines is to attract other molecules into the tissue to help clean up the wound and kill off invading bacteria. And unfortunately the way these responding molecules work produces oxidative damage.'

Given all the bad things senescent cells do, the obvious question is why, when it becomes clear that a dividing cell is irreparably damaged, does the tumour suppressor p53 not automatically trigger the suicide mechanism, apoptosis, and consign the cell to the recycling bin rather than inducing senescence? 'Exactly!' exclaimed Campisi when I visited her at the Buck Institute in California to talk about her research. 'That question bothered me for a long time.'

There were hints in the literature that perhaps senescent cells had some positive attributes, too, and so Campisi and her colleagues took a closer look at them. They found that as well as collagen-chewing molecules and inflammatory cytokines, the secretory substances also include growth factors – molecules that encourage repair and regeneration of tissue. They already knew that in order to heal a wound, our bodies need to mount an inflammatory response (that's why, when you cut yourself, the flesh around the cut goes pink and hot for a while). To test a hunch that maybe senescent cells are actively involved in wound healing, Campisi's team created a transgenic mouse in which these cells expressed a protein that made them glow so they were easily identifiable among other cells in the body. They then made small wounds on the backs of their lab mice and found that, sure enough, senescent cells were clustered at the site of the wounds. When they eliminated the senescent cells, the wounds were very much slower to heal.

Researchers elsewhere had discovered exactly what type of growth factor the cells at the site of skin wounds were making. To test the significance of their findings they cleared out the senescent cells around skin wounds on their experimental mice; they noted that the wounds were struggling to heal – clearly lacking something crucial to the process – and then applied the missing growth factor that would have been supplied by the senescent cells as an ointment instead. The wounds healed just fine, says Campisi.

So far the scientists have focused on senescent cells at the site of skin wounds. However, senescent cells are not all the same; they vary according to where in the body they're found and the tissues or organs they come from. Campisi suspects that those in, say, the liver or kidney will secrete substances important for healing injuries in these organs too. There's tantalising evidence for this. Scientists at University College London (UCL), led by Maximina Yun, working with salamanders to try to discover how they are able to regenerate whole limbs that get lopped off, have found that senescent cells accumulate in the buds of the regenerating limbs but are efficiently removed later in the process. They believe the senescent cells are playing an active role in the regeneration – a hunch supported by their very recent findings that these cells are necessary for the normal development of amphibian embryos: eliminate them and the hatchlings have defects. Researchers have found the same thing with mice, too – that senescent cells help remodel tissues during development. 'It's this double-edged sword again,' comments Campisi. Senescent cells, it seems, are needed to promote tissue repair and for healthy development. 'But you don't want them sticking around.'

That's the crucial point, because they *do* stick around as we age. No one yet knows why or for how long, but they are the cause of chronic, low-grade inflammation as the persistent secretions keep up the signals to the immune system. 'I don't know if you've ever seen the way immune cells get into tissues?' said Lynne Cox, Associate Professor of Biochemistry at Oxford University in the UK, who shares with Judith Campisi a special interest in senescent cells. 'They actually break a path between cells. If they're crossing through a blood vessel, for instance, they break apart the structure and climb through the vessel – it's really cool!'

Normally, a white blood cell in a young person will home in on a site of damage, and take a fairly direct route, Cox explained. But researchers in Birmingham, UK, have shown that these immune-system cells in old people seem to lose their sense of direction and go zigzagging through the tissue towards the site of injury, wreaking havoc along the way. 'So, in an old person,' commented Cox, 'not only do you have more inflammation because you've got more senescent cells kicking out these inflammatory cytokines, but your immune cells are causing damage as they get to the site of injury.'

Like Judy Campisi in the US, Lynne Cox came to cell senescence from the cancer research field. A petite, vivacious woman, she sweeps you into her world with her bright, bubbly enthusiasm. Cox has been interested in science for as long as she can remember, she told me when I visited her in Oxford. 'My mother tells me that I was always doing "speriments", even before I could say the word,' she laughs.

'When I was still at primary school I was sending her off to the local toy shop to buy cobalt chloride so I could paint weather pictures. Because this chemical is blue, and when it gets damp it goes pink, and you can actually predict the weather from these pictures. Heaven only knows where I got that from!' She remembers, too, an explosion and a lot of broken glass in the living room of the family home where she had been messing around with ammonia as a very young kid.

Cox began her professional career in the Dundee lab of David Lane, working on the tumour suppressor p53, which he had co-discovered in 1979. The focus of her early research was how the gene 'supervises' the replication of cells. p53 is a master switch at the heart of a network. To do its work it receives signals from damaged cells and responds by switching on a suite of genes to execute its programme. The gene Cox was most interested in was one called p21, which is responsible for taking a damaged cell down the path of 'arrested division' rather than suicide. If the DNA in a cell is not badly damaged, p53 will give it a short burst of p21 activity, which will give the cell time to be repaired before going on to complete the process of division. But if the DNA is badly damaged, p53 will give it a strong blast of p21 that will consign it to permanent arrest, or senescence.

There are two schools of thought about how to prevent the damage done by senescent cells. One strategy is to develop agents known as senolytics that will kill the cells, sending them off for recycling. The other is to rejuvenate the cells so that they function normally again. There are pros and cons to each approach. In recent years, a number of labs around the world, including Judy Campisi's in the US, have managed to kill senescent cells in various sites in mice. 'When you do that under disease settings you can show that, if you force the cells to die early on, the mice don't develop a severe disease,' says Campisi. 'And in some cases, if you force the cells to die even after the disease has developed, you can get some improvement. It depends on the tissue, and it depends on the disease.'

Campisi's lab has looked at the effects on arthritic joints in mice and found that after a few weeks or months the damaged joints showed signs of repair and the animals were no longer hobbling. And in 2016 researchers at the US's Mayo Clinic, also working with mice, announced that they had managed, with repeated treatments of senolytic drugs, to clear or prevent the build-up of calcium in blood vessels that leads to cardiovascular diseases. They hope this will translate into new methods of treatment for people with hardening of the arteries to replace or complement the only option available today, which is surgery.

Another elimination strategy, says Oxford University's Lynne Cox, is to force senescent cells to self-destruct by tinkering with the mechanism that normally makes them resistant to suicide. This has been done on middle-aged mice and the effect has been striking, she says. But the worry with senolytics in general is that no one knows what would happen if they were used on really old animals – or people – whose bodies are riddled with senescent cells ... The mind boggles; such a strategy could lead to catastrophe.

There is always the risk, too, that, in trying to limit the damage, you will be tampering with the positive things – some vitally important, and some yet to be discovered – that senescent cells do. Take skin cells, for example. They form the outer covering of our bodies and the interface with our environment, both internally and externally. As such they are especially vulnerable to damage, and they divide rapidly for maintenance purposes. One good reason why skin cells are primed to senesce rather than to commit suicide when the damage signal goes off is that we cannot afford to lose too many of the cells that hold us together.

So what of the second option described by Cox: simply reversing senescence so that cells start functioning normally again? Researchers have managed to do this in human cells in lab dishes by switching off p53 and p21, which are responsible not only for driving cells into senescence but for maintaining that 'arrested' state. By manipulating these genes, the scientists were able to override the Hayflick limit and keep cells going with critically short telomeres until they accumulated so much damage that they went into crisis. But this strategy is essentially switching off the tumour-suppressor system, and the big danger is that an unrestrained damaged cell might turn cancerous.

This is the danger, too, with yet another strategy for rejuvenating senescent cells, which is to turn on telomerase (not active in most body cells, you'll remember) and rebuild the truncated telomeres. But this is an approach that 'worries the socks off me,' says Cox. 'If you turn on telomerase in middle-aged mice, they do rejuvenate. The muscle mass increases; the gut shows improvement; I think even their brain size increased [in the experiments]. But if the mice had a precancerous tumour, it became very aggressive very quickly.' Given that precancerous tumours, kept in check by healthy tissues and efficient immune systems, are a common occurrence in us humans, Cox's alarm about telomerase is well placed.

She and her team are investigating an altogether different approach to rejuvenation that uses a drug to target a central component in the cellular machinery. This component, an enzyme called TOR, helps cells produce the proteins that carry out all the tasks in the body, and it's particularly busy in fast-dividing cells, helping them to bulk up in preparation for division. If TOR is working too long or too hard it will drive a cell to senesce. But if you inhibit TOR, you slow down protein production, and you turn on the housekeeping programme that recycles all the worn-out components; senescent cells become younger again, says Cox. They have been trying out the strategy in lab dishes. 'We've got some drugs that can actually take a deeply senescent cell and make it look young again, and behave young again, and carry on proliferating … I've got some in the incubator if you want to see them.'

I most certainly did. There's nothing like looking at the real thing to make sense of all the talk, so I followed Cox down the corridor, donned a lab coat and peered down a powerful microscope at a series of slides she pulled from the incubator. There were normal skin cells from a young person, and then senescent cells from the skin of an older person. The latter did indeed look old and raggedy by comparison, reminding me of the way in which an egg well past its sell-by date broken into a frying pan will splurge across the surface, rather than holding its shape firmly as a new-laid egg will do. And then there were the rejuvenated cells that had miraculously perked up.

By using a blue dye, developed by Judy Campisi's group, that stains only senescent cells, Cox and her team were able to follow the fate of the cells as they drugged them. Before they administered the drug, 65 per cent of the population of skin cells in their culture stained blue, indicating that they were senescent. After a week of treatment, only 15 per cent were still blue, and the rest were happily dividing like young things again. Interestingly, when they withdrew the drug, the senescent cells gradually reverted to their aged selves. What's more, the scientists were able to toggle them back and forth between old and young, and to keep many of them going for scores of generations – way beyond the normal cellular limits.

The drug used by Cox and her team is modelled on one that has hugely excited the gerontology community since 2009, when scientists at the National Institutes of Health in the US, during a prospecting search for anti-ageing compounds, found that it extended the average lifespan of mice by 12 per cent, while also significantly improving their health. Rapamycin is already in the medicine cupboard; it is an immune modulator and is used traditionally to dampen down rejection of transplanted organs. The drug, produced by a soil bacterium, *Streptomyces hygroscopicus*, found on Easter Island, was isolated for the first time in 1972. It was originally developed as an antifungal treatment, and named rapamycin after the native name for Easter Island, Rapa Nui. Rapamycin is being used today in labs across the world to test its multiple effects on ageing processes.

In 2013, researchers at the Buck Institute, for example, reported that elderly mice suffering from the same kind of cardiac problems as we humans suffer in old age – enlarged hearts with thickened walls, and heightened blood pressure – showed marked improvements in cardiac function and general fitness after just three months of treatment with the drug. By comparison, their untreated lab mates had declined further.

The drawback with rapamycin, however, is that it can have a range of more or less uncomfortable and serious side effects, ranging from constipation and swollen ankles to abnormally high levels of cholesterol and sugar in the blood, and increased risk of type 2 diabetes. By suppressing the immune system – the quality that makes it so useful for avoiding rejection of transplanted kidneys – it can also leave people vulnerable to infections.

In 2013 they didn't know how the drug worked to produce the effect they had seen on the mice's hearts, except that it affected the TOR network, that central bit of the cellular machinery that Cox and her team targeted to rejuvenate the senescent cells I saw in their lab. (In fact, TOR stands for 'target of rapamycin'). But further investigation revealed that it suppresses or alters the secretions from the senescent cells that cause so much havoc in the tissues. It also appears to switch on a natural recycling programme known as autophagy that dismantles damaged cells, salvaging the components to make building blocks for new cells to replace them.

In 2015, Campisi's team, in collaboration with others at the Buck Institute and elsewhere, found that by manipulating the dosage of rapamycin and giving it intermittently they could selectively block the element in the secretions that causes inflammation – the big bugbear in ageing – without blocking the secretory elements necessary for wound healing. The inflammation response, they found, is so intricate and complex that it takes the cells a long time to fire up again, once it has been interrupted – hence the effectiveness of intermittent dosing. Importantly, such a dosing regimen limits the risk of side effects. 'We think this could provide a paradigm shift in the treatment of age-related disease, including cancer,' said Campisi, when the results of their study were announced. 'Imagine the possibility of taking a pill for a few days or weeks every few years, as opposed to taking something with side effects every day for the rest of your life. It's a new way of looking at how we could deal with age-related maladies.'

That same year, researchers at Washington State University started enrolling people's pet dogs in a trial with rapamycin – reasoning that the promise of slowing ageing and improving the health of man's best friend would capture the public imagination in a way that work with mice and flies and worms just doesn't, and hopefully garner popular support for ageing research. Only a handful of middle-aged big dogs – which generally have shorter life expectancy than little dogs – have been recruited, and the aim of this particular trial is to check that the drug is safe. Which it has proven to be thus far. The next phase will test longer-term use of the drug and its effectiveness at delaying the many maladies of old age.

Animal models – worms, flies, mice and, yes, now dogs – have taught us a huge amount about senescent cells and how we might manipulate them. But what about humans? For insights into what senescent cells do to us, we can look to people with one of the progeroid (or premature ageing) conditions.

CHAPTER FIVE

Old before their time

I met Mark Jones* and his mother Pat over a cup of tea in the rather fusty grandeur of an old London hotel on a cold January afternoon. Now in his late thirties, Mark is a small man with neat, bird-like features and little to mark him out obviously apart from a high-pitched and rather hoarse voice, as though he's suffering from laryngitis. But having been tiny at birth – he weighed about 1.8kg (under four pounds), though he was not premature – he never caught up with his peers; he didn't experience the usual growth spurt at puberty and stands at 160cm (5ft 3in) today. Mark has suffered all his life from skin problems, and stiffness and pain in his tendons and joints, which occasionally swell up and make it hard to get out of bed. 'I've had a permanent pair of crutches for 15 years, and every now and then one of my legs plays up and I have to use them, because it's really, really painful and I can't bend it or move it,' he told me.

These episodes are tough for a man who has always loved sports, and he picks up the traces of everyday life again as soon as his body will allow. He ran the London marathon for the second time when he was 31, but an old ruptured tendon in his knee played up and he was forced to walk half the course. He has had cataracts removed from his eyes, a hip replaced, and he has to look after his skin carefully, especially on his feet and hands, which feel tight and are super-sensitive to the cold.

---

* Not his real name. Changed to protect the privacy of himself and his family.

For years Mark would visit his doctor when things flared up, and he would occasionally be referred to specialists for tests, but they could never find anything that might be causing his problems. It might be his sporting activities, some suggested lamely. Apart from the fact that diagnosis is a fine art anyway and Mark's is a rare condition, the trouble was that everyone he saw treated him symptom by symptom. It wasn't until he was offered a routine health check at work that anyone stood back and looked at the bigger picture, and he began to get answers to the questions that had been nagging him and his family.

Mark was asked to list all the conditions for which he had sought treatment over the years and says, 'The person doing my health check was fascinated! She said, "Oh there's something more to this … Has anyone ever tried pulling things together to reach a conclusion?"' He was referred to a rheumatologist who took a full history, examined him carefully and suggested his patient might be suffering from Werner's syndrome. The rheumatologist advised a DNA test, and a sample of Mark's blood was sent for genetic testing in England, and then to the Netherlands for final analysis. That was October 2015, and Mark finally got confirmation of the diagnosis of Werner's, a premature ageing syndrome, in April 2016.

'It's a funny scenario, because you think, "Okay, now they've got a name for it; that's good, because they can deal with it,"' he reflects. 'But then you look into it and realise it's not like that … It's not like you can throw a few pills and sort this out. So in a way it was a concern, but we just thought, "Okay, we'll see how it pans out."' Trawling the internet for information just added to his concern as he read of survival rates (average life expectancy 46 years), the range of conditions he could experience (including osteoporosis from loss of bone density, cardiovascular diseases, early-onset cancer, and skin, muscle and tendon atrophy) and the effects it might have on his appearance. People with Werner's tend to experience early greying of the hair and baldness, and develop pinched facial features due to loss of fat below the skin, and stick-like limbs with a tubby trunk. 'You always look at the worst-case scenario,' comments Mark. 'It looked like I could have only eight years left. And I'm thinking: I haven't been to South America yet; I want to walk on the Great Wall of China ...'

'One of the key features of Werner's syndrome,' says Richard Faragher, who studies the condition at the University of Brighton, 'is greatly compromised proliferative capacity of the cells. Leonard Hayflick, in the 1960s, demonstrated that normal human cells have a finite limit to proliferation: the population only doubles in number about 50 times. With Werner's syndrome, more than 90 per cent of cells in culture do less than 20 population doublings. Now this is a vast difference in cell proliferation.' Especially when you think of it in terms of cell numbers, which expand exponentially, says Faragher. The total number of cells available for repair and maintenance of a tissue is two to the power of the number of doublings. 'So normally that would be two to the power of 50; in people with Werner's syndrome it's two to the power of 20. That's a huge difference. It's not a twofold difference, it's a 30 *power* difference – the equivalent, I've worked out, of around 1,125kg [2,480lb] of cells over a lifetime.'

The overwhelming evidence, Faragher believes, is that most of the clinical features seen in people with Werner's syndrome are the result of the accumulation of senescent cells, as they outstrip the capacity of the immune system – itself getting less efficient with age – to clear them.

Faragher has been studying Werner's since he did his PhD on the topic in the early 1990s. His journey into the field of ageing research had begun when, as a six-year-old, he had seen a BBC documentary on the family's old black and white TV that talked about the Hayflick limit. 'That documentary,' he recalls with a wide grin, 'communicated what all science documentaries do, which is on the one hand, "Isn't it fascinating and wow!", and on the other, "Goodnight, children, the scientists are on the case. That cure for cancer is coming right along …!"' Faragher was hooked; from then on a scientist is what he wanted to be. 'But what I realise is that I also ticked many of the social boxes that indicate people will go into a career in geriatrics,' he says. 'Because my parents … we had lost our home; we lived with my grandparents, and living with older people is a very, very strong predictor of not seeing them as some kind of twisted mutant from hell. They're real people; and I was extraordinarily close to my grandmother, which was very, very pleasant.'

When Faragher graduated with a biology degree from Imperial College London in the 1980s, the field of gerontology barely existed; the literature on the biology of ageing was extremely patchy. But rummaging through the shelves of London's famous bookshop, Foyles, he discovered a little book called *Studies in Biology 151: Ageing*, which he pounced upon excitedly to read on a long coach trip to the coast to visit a friend. In it was a short passage on Hutchinson–Gilford syndrome, a disease that causes premature ageing in young children, and Werner's, which tends not to manifest itself until around puberty. What particularly intrigued Faragher was the fact that each was thought to be caused by a single-gene mutation, though nobody knew then which genes were involved. This was a time when scientists searching for the mutant genes behind a number of serious inherited conditions were registering success with diseases like cystic fibrosis and the muscle-wasting Duchenne muscular dystrophy. Discovering genes that did this or that was in the air, and very soon Faragher had decided that what he really wanted to do was study Werner's syndrome and find the gene that when mutated gives 'humans multiple features that look a hell of a lot like ageing'.

In the end he was pipped to the post – Faragher finished his PhD in 1994, and the Werner's gene was discovered by George Martin and colleagues at the University of Washington in Seattle in 1996. Named simply WRN, it was found to be the recipe for an enzyme that unwinds DNA during the process of cell division, when the genome is being copied. But what Faragher did discover during research for his doctoral thesis was that the reason Werner's cells have such a shortened lifespan is that they are three to five times more likely to become senescent than normal human cells. We saw in an earlier chapter how telomeres measure out the Hayflick limit – the normal lifespan of cells – as they shorten with each division. In Werner's cells the telomeres seem to shorten at a faster rate than normal, and to trigger the alarm signals earlier – or at a longer length – than they would in normal cells.

The link with telomeres – and hence also with telomerase, the enzyme that replenishes telomeres in some tissues – has shed light on another perplexing feature of Werner's syndrome: that unlike normal ageing, it doesn't involve the whole body. 'The immune system and neurons do not appear to be affected,' says Faragher. 'So what you have is an interesting mosaic where some tissues are affected very, very severely and some tissues aren't affected at all. So okay, what the hell is going on? In essence it turns out that tissues where telomerase is turned on are quite protected against the effect of Werner's; tissues that are telomerase-negative aren't protected, and they're badly affected, typically.'

But telomeres are only half the story. The abnormal behaviour of these protective caps of chromosomes is not enough on its own to explain the exceptional rate at which Werner's cells senesce. We saw in the last chapter how damage of all sorts to DNA – from UV radiation from sunlight to the activity of free radicals or toxic chemicals – can stop a cell in its tracks. It turns out that Werner's cells are more sensitive than normal to damage signals from any source, and more ready to slam on the brakes. But why?

Lynne Cox, the lady who showed me the senescent cells under a microscope, discovered this crucial piece of the jigsaw quite by chance. She has been interested in the mechanism of cell division and replication since her earliest days working in cancer research in Dundee, Scotland. Her special focus is a little piece of the machinery that slides along the ribbon of DNA, orchestrating the production of the two new strands by pulling in proteins to perform this task and that, and clearing up the debris as it goes.

Attending a conference on cell replication at Cold Spring Harbor Laboratories near New York in 1996, Cox was strolling around the posters that research groups are invited to pin up on display boards on the fringes of these talk shops to flag up interesting findings. She came across one telling of the discovery of the Werner's gene and including the observation that people with a mutated WRN gene get old quickly. No explanation; no theory as to why; just a simple observation. Typically, hundreds of posters are displayed at big scientific gatherings, and Cox could so easily have missed this one. But pausing before it, she was quickly intrigued. She knew by heart the genetic sequences of the proteins that slotted into her little piece of kit – the replication 'platform' – and WRN seemed to carry a familiar code. Could it be one of the proteins that slot into the platform to carry out a crucial task in replication?

Back in her lab in Oxford, Cox tested her hunch and found she was right. Basically, this is how it works: in the normal course of events, when the replication platform trundling along the DNA ribbon hits a patch of damage, it stops and sends a signal out for WRN, which attaches itself to the platform. WRN has a little pair of molecular scissors which it uses to snip out the damaged segment and then it falls away, allowing the machine to carry on along the DNA ribbon. But if the WRN gene is faulty, it doesn't answer the call and the whole process stalls, with disastrous consequences for the cell.

Replication platforms 'are a bit like bicycles,' explains Richard Faragher, 'they're very, very stable provided they keep moving. If they're stopped for too long the whole thing collapses. DNA repair enzymes move in to sort out the mess and the whole structure is typically resolved by deletion. So you can lose big chunks of stuff.' What you're left with is cells with lots of little chromosome breaks and aberrations that alert the anticancer mechanisms and the cells 'crash out into senescence'.

It's the chronic inflammation set up by these cells that is the cause of Mark Jones's joint, tendon and back pains, and of his skin problems. Inflammation is involved, too, in the diabetes he was recently diagnosed with, and in most other conditions he and his fellow Werner's patients suffer. This is 'inflammaging' writ large, and the small group of gerontologists who are focused on Werner's syndrome see it as a model of how ageing proceeds – at a more leisurely pace – in the rest of us. They have worked out how the chaos in a Werner's cell triggers senescence – how it signals its distress and sets off the cascade of events that bring it to a permanent halt, when it starts doing bad things. At the head of this cascade is a protein that goes by the prosaic name of p38 MAPK*. This controls the activity of players downstream and promotes the synthesis of the inflammatory cytokines that we met in the last chapter – the molecules

* The initials stand for p38 mitogen-activated protein kinase

that fire up the immune system and keep it ticking over indefinitely to cause chronic inflammation.

p38 MAPK's role in inflammatory conditions such as rheumatoid arthritis, psoriasis* and Crohn's disease† has been known for a long time, and a number of big drug companies have been working their socks off to find safe and effective ways of blocking it as a treatment for these diseases. David Kipling and colleagues at Cardiff University, Wales, however, were the first to stand back and see the bigger picture: that p38 MAPK is not just involved in promoting inflammation, but is a key player in the whole business of responding to stress, and therefore in shutting cells down, causing senescence. This is the case in people with Werner's syndrome at least, and very probably in all of us.

Proving the hunch that this was the case meant working with agents that block p38 MAPK to see what effect they had. Because of the high stakes involved in developing a drug with such huge market potential, Big Pharma plays its cards close to its chest, and the companies involved were reluctant at that time to share their compounds. So Kipling turned to Mark Bagley, now Professor of Organic Chemistry at Sussex University, for help.

'One of the problems,' said Bagley when I visited him at his lab on the leafy campus on the outskirts of Brighton, 'is that p38 MAPK is a protein that you can't knock out – it's absolutely critical to a lot of processes. What David was interested in trying to do was to prove that this

---

* A skin condition that causes red, flaky, crusty and itchy patches of skin covered with silvery scales. These patches normally appear on the elbows, knees, scalp and lower back, but can appear elsewhere too.

† A type of inflammatory bowel disease that may affect any part of the gastrointestinal tract from mouth to anus.

protein was essential, and that it was involved in Werner's syndrome pathology and so was involved in accelerated ageing. The question was: does a similar accelerated ageing happen in normal people, during certain events? Does stress trigger ageing of cells prematurely, before the telomeres have been eroded?' It's an important question: who hasn't wondered, watching the swift transformation of well-known figures like Tony Blair and Barack Obama from fresh-faced and youthful when they took office as heads of the British and US governments respectively, to careworn, grey-haired and increasingly haggard men by the time they left office a few years later, whether the stress of the job was to blame?

'I was in Cardiff at the time, and the medical school and traditional university had just merged, so new opportunities to collaborate had opened up,' said Bagley, a big, smiley man with a mane of ginger hair and a shaggy beard. 'It was an email out of the blue. So David and I met for a coffee ... When scientists collaborate it's very important that you get on, that you can talk, and we got on brilliantly!' Bagley agreed that, if Kipling could find the funds to employ enough pairs of hands in his chemistry lab, he would work on compounds that block p38 MAPK.

Developing a compound that was specific to this particular protein – one that could silence just a single note in a veritable cacophony of communication – was painstaking indeed. Because p38 MAPK is critical to so many cellular processes, Bagley and his team could not simply knock the protein out without doing huge collateral damage. The Werner's cells the scientists were working with were precious: with so few people known to be affected, the supply was extremely limited. Those cells they cultured in their lab dishes were horribly stressed from the processes involved. They were all at different stages in their lifespans. And they were genetically unique, having come from different individual patients, so no two cell lines behaved identically, making it a real challenge for the cell biologist to tease out what was happening.

'We made, I think, about 8 to 10 different chemical inhibitors, all of which were first targeting p38 MAPK, and then moving downstream from p38 MAPK to see if we could avoid some of the toxic effects of taking out this protein. Then we tried to move to other signalling pathways to see if they were also involved. The studies have probably taken the best part of 15 years,' says Bagley.

The conclusion of their research was that p38 MAPK – as a key player in the response to stress – is directly involved in the premature ageing of people with Werner's syndrome. Their results also suggest the likelihood that stress can, under certain circumstances, accelerate ageing in any of us by triggering senescence in our cells even before their time has run out on the telomere clock. But besides giving vital insights into how ageing occurs, the collaboration between the Kipling and Bagley labs could offer immediate practical help to people living with Werner's syndrome.

More people have been diagnosed with the condition in Japan than anywhere else on Earth, and in that country it is thought to be caused by a 'founder mutation' – that is, a mutation that was introduced into the population by a single individual who is a common ancestor of many of those carrying the mutant gene today. Richard Faragher has been working closely with a group of clinicians caring for patients in Japan and says, 'I went along to see for myself the actual problems these guys face, and it's pretty terrible. Their life expectancy has gone up, but their quality of life isn't good.'

A major issue is the failure of wounds to heal. Many people develop ulcers, typically on their feet and ankles, caused by atrophy of the dermal layer (the middle of three layers) of skin so that it tears easily. The ulcers eat their way down to the bone. Attempts to cover them with skin grafts from the patient's back or arms – painful operations in their own right – usually prove futile, so that sufferers are faced eventually with amputations of their feet, and life in a wheelchair.

In the course of their research, Kipling and Bagley found that Werner's syndrome skin cells treated with p38 MAPK inhibitors reverted to normal: their super-sensitivity to DNA damage disappeared and they lived out a normal lifespan. Faragher is keen to try the compounds on patients' ulcers. A number of p38 MAPK inhibitors developed by Big Pharma are already available for use in the clinic and have proved effective, taken as pills, in trials with patients with rheumatoid arthritis. But they are not yet in the medicine cupboard because they turned out to be toxic (particularly to the liver) with long-term use. Faragher's intention, working with one of the consultants in Japan, is to try applying one of these compounds as an ointment around the edges of the ulcers, and then discontinuing treatment as soon as the ulcer closes up.

'I very much want to do that,' he says. 'I think that after 20 years ... the patients have helped us and I would really like to see a practical outcome. Because what are we doing? We're normalising cellular growth. We're seeing the problems caused by cellular senescence in Werner's patients, and I'd like to see that mechanism sorted.'

But when you talk about the possibility of boosting the replicative capacity of Werner's syndrome skin cells, he comments with obvious frustration, 'There are always people who say, "What about the cancer risk?" My response to that is: which would you rather have? A bilateral amputation or a risk, *maybe*, of cancer? Your call, mate ... So I'm very hopeful that we will finally get something done.'

A lot can be learnt about ageing from studying cells in a dish. But sooner or later theories have to be tested in living organisms because, in life, cells talk to one another the whole time, and influence one another's behaviour. Werner's syndrome and other premature ageing conditions are very valuable in offering insights, and a huge advantage is that it's human biology, not the biology of proxy organisms, that is under the microscope. But premature ageing conditions do have certain limitations. One is that they are caused by single-gene mutations, whereas normal ageing is the product of multiple genes often just doing what they are supposed to do. Another limitation is that none of the premature ageing syndromes affects the whole body.

An important point to make about the limitations of Werner's syndrome in particular as a model of ageing is that, of course, not all cells in our bodies divide constantly. Some, such as liver and kidney cells, do so only rarely when kicked into activity to repair damage. Others, including heart, red blood, brain and nerve cells, are classified as non-dividing. Cellular senescence, therefore, could never be the sole mechanism responsible for ageing – even if a sole mechanism behind such a complex phenomenon were likely. In future chapters we'll look at some of the other hypotheses and other mechanisms. But right now I'm going to take a break from the hard sleuthing for a while to introduce you to some of the most important non-human actors in the drama of ageing research.

CHAPTER SIX

Ming the Mollusc and other models

'*Fruit flies*?' said my partner incredulously. 'What, those teeny things we see flitting round our food waste bin?' I had been telling him how much has been learnt about the mechanics of ageing – and of age-related diseases, from dementia, cancer and heart failure to diabetes, inflammation and failing eyesight – by studying the fruit fly (or *Drosophila melanogaster* to give it its scientific name). It was a good question: just how do scientists work with such tiny creatures to uncover the secrets of life and death? I was curious to find out.

Mar Carmena has been a fly biologist for nearly 30 years. Her principal focus has been teasing out the mechanics of cell division, a process that enables a fertilised egg to grow from a single cell into a mature adult by faithfully following a fiendishly complex programme of development. Carmena first encountered flies as model organisms as an undergraduate in her native Spain, where she was helping out in the lab in exchange for financial assistance with her studies at the Universidad Autónoma de Madrid. 'It was a very exciting time,' she told me in her soft Spanish accent. 'It was the late eighties and they had just discovered that many of the genes that determine the body patterning in flies were conserved in humans. So they were basically the same biological processes behind body development in both species.'

Carmena continued working with flies as a postdoc in Dundee, in the lab of Professor David Glover, one of the driving forces in the worldwide collaborative effort to sequence the genome of *Drosophila*. Today she heads a *Drosophila* research project at the University of Edinburgh, where I met her in her office with its awesome views out over Arthur's Seat – the grass-clad extinct volcano that stands tall at the heart of the city. I wanted to take a close look at her model organisms. Carmena and Emma Peat, who's in charge of breeding and caring for the fly population, took me to the little room where their charges are kept, segregated by generation and genetic background, in hundreds of stoppered flasks. The windowless room has a sickly malt-vinegary smell from the yeast and sugar mixture used to feed the flies. There's a net curtain across the entrance as an extra precaution against any genetically modified critters getting out into the wide world.

The flasks teem with what look like the tiny iron filings we used in school to learn about magnetism. But once under the microscope, sedated with puffs of carbon dioxide and magnified up to fiftyfold, the fly's structure becomes awesomely clear. The head, thorax and abdomen are nicely distinct; you can pick out the individual cells in the orangey-colored compound eyes that look like pincushions with their tiny spikes; the hairs on its body and legs stand out like thorns on a cactus; you can see the internal organs working, and even determine its sex. But what is equally astonishing – and particularly important for researchers – is how your mind adjusts to the size of the image before your eyes, so that, with a bit of practice, you can touch and dissect the animal with the whisker-fine specialist instruments almost without inhibition, as though it were a regular size.

The first person to suggest the fruit fly as a model organism for genetic research was the entomologist Charles Woodworth, who, at the end of the nineteenth century, bred them for the first time in his lab at Harvard. But the man who really propelled flies into the mainstream of scientific research was Thomas Hunt Morgan, an embryologist who started using *Drosophila* in 1908 to study the mechanisms of inheritance. Fruit flies appealed to him as a model because they breed fast, have a short life cycle and are cheap and easy to manage.

Born in 1866 in Kentucky into a prosperous Southern planter family, nephew of a Confederate General and great-grandson of the man who wrote America's national anthem, *The Star-Spangled Banner*, Morgan grew up a nature lover – a collector of birds' eggs and fossils. He began studying science at college at the age of 16, and had a PhD in zoology from Johns Hopkins University by the age of 24.

At Columbia University, New York City, in the early 1900s, he set up the Fly Room. A space just 5m (16ft) wide by 7m (23ft) long, crammed with eight desks and hundreds of glass flasks filled with flies, a bunch of overripe bananas (fly food) dangling from the ceiling, and a pervasive smell of fermentation, the Fly Room was to become world famous. Here Morgan confirmed the chromosome theory of heredity which suggested that genes – the units of inheritance that had only very recently been named – are carried on chromosomes like beads on a string. He published his work, *The Mechanism of Mendelian Heredity*, in 1915. It was to earn him a Nobel Prize in 1933 and the title of 'father of genetics' – though this honorific is just as often attributed to Gregor Mendel, the Augustinian monk/scientist whose abstract theory it was that Morgan had confirmed.

In 1856 Mendel had been given permission to carry out research in the large experimental garden of his monastery in Brno, then part of the Austrian Empire, today in the Czech Republic. Interested in science since his college days, Mendel began to work with pea plants, breeding and cross-breeding tens of thousands of them for certain traits, such as height and colour, to try to tease out the rules of inheritance. He described his findings and expounded his theory in two lectures given at a scientific conference in 1865.

But there's a rich irony here, for Morgan had begun his fly studies as a sceptic: he didn't accept Mendel's suggestion of 'invisible factors' driving inheritance – nor yet Charles Darwin's theory that the mechanism of evolution was natural selection – and he had set out to challenge the two scientists' hypotheses. But what he witnessed in the Fly Room provided clear, experimental evidence of these abstract ideas and changed Morgan's mind decisively. The genes, he realised in what must have been a heady period of discovery in that fusty little room, were in fact the unifying factors behind the three questions that most intrigued him: what is responsible for carrying physical traits from one generation to the next; what is the mechanism behind evolution; and how does an egg develop into an embryo and then into an adult creature?

The Fly Room, which existed from 1910 to 1928, was enormously influential. The research pioneered there helped make the fly one of the most widely used model organisms in biology, and one that produced several more Nobel Prize winners over the years. Notable among them was Hermann J. Muller of Edinburgh University, who discovered the mutagenic effects of radiation on DNA, winning the Prize for Medicine in 1946.

By the year 2000, the fly genome had been fully sequenced and found to be made up of around 13,600 genes, carried on just four chromosomes. And when the human genome was finally sequenced in 2003, it revealed that 60 per cent of the fly's genes occur in us, too – a fact that boggles the mind, for it means that these genes have been conserved in both species over unimaginable eons of time and down countless generations from our common ancestor who lived some 3.5 billion years ago. We now know that around 75 per cent of genes known to be involved in human diseases are also found in *Drosophila.*

The creature's relatively simple genome is easy to manipulate to mimic desired effects, and the facts that the fly reproduces fast and prolifically, is extremely cheap to maintain and doesn't invite attention from the animal rights brigade make it attractive to biologists of all sorts. Over the years a huge amount of information has been gathered about the fruit fly, and a comprehensive toolbox of techniques for tinkering with its genome put together. Today, researchers can obtain off-the-shelf mutants of almost every sort from fly-stock centres in the US and Europe, and have access to the FlyBase website with its wealth of knowledge about how the fly works and how to carry out experiments.

With just 100,000 neurons compared with our 100 billion, the fly's brain, too, provides an appealingly simplified system for exploring basic principles of brain function and disease processes, and for testing theories. In fact, neuroscientists who first turned to the fly as a simple model found that its brain is remarkably similar to that of mammals, including us. It has a blood–brain barrier designed to protect the brain from harmful stuff circulating in the rest of the body; a complex central nervous system; and organisational features a lot like ours.

And isolating the brain can be a piece of cake. When it gets to the third larval stage, says Mar Carmena, the fly is basically a sac of tissue, most of which will be junked and recycled when it goes into the cocoon to metamorphose into an adult fly. But within that sac, the brain and central nervous system of the adult are already formed. 'So we can actually pick up a larva and treat it, for instance, with your cancer drug of choice, and see the effect in the nervous system … I mean obviously there are limitations to the conclusions you can make, but in the basic processes of cell division and differentiation you can actually start seeing if there's going to be a problem.'

Though it doesn't normally suffer a fly version of dementia, *Drosophila* does share with us the gene involved in the hereditary form of Alzheimer's disease. Building on this, researchers have made transgenic flies that can mimic various aspects of Alzheimer's, including the build-up of protein plaques in the brain, resulting in loss of memory (yes, believe it or not, flies do have memories – and ones that can be measured), locomotor defects, learning difficulties and early death. These transgenic flies – and also creatures 'designed' to mimic a range of other brain diseases – have been used to study the processes of neurodegeneration of all sorts and to test ideas for intervention. Some of the mutants created by researchers have wonderfully florid descriptive names. The 'drop dead' mutant, for instance, dies prematurely with a degenerate brain; the 'spongecake' mutant develops pathology very similar to Creutzfeldt–Jakob (or mad cow) disease; and the 'Swiss cheese' mutant, as it ages, shows pathology reminiscent of motor neurone disease.

A lot can be read by studying the fly's compound eyes, which are like a window into its brain. Made up of around 800 individual units rich with nerve cells, the eyes are quick to show signs of degeneration as the surfaces roughen and shrink and the cells lose their pigmentation. Very recently the fly has been adopted as a model for studying the heart, too – specifically, how the electromechanical signalling system works to maintain the precise choreography of the heartbeat across a lifetime, and how that begins to falter with age. This is one of the least well-understood aspects of heart function, because by its nature it must be observed in still-living creatures, as they age naturally, and there are no easy models. Thus the fly is a pioneer, and these studies have been made possible by new imaging techniques developed to watch its heart beating in real time, as it ages across a matter of days rather than years. I have seen a video recording of this, and it is awesome, especially when I remind myself again of that tiny pest circling my fruit dish.

There are, of course, big differences between fly and mammal hearts. Flies have an 'open circulatory system', which means they don't have blood vessels, veins or arteries, and the architecture of their hearts is much simpler than ours. Nevertheless, the model has shed light specifically on what happens when the heart rests between contractions and just how this changes as we age.

Another stalwart of biological and genetic research is the tiny worm we met in Chapter 1, *Caenorhabditis elegans* – 'just about the size of a comma in a sentence', according to molecular biologist Cynthia Kenyon – which lives in soil, swimming in the film of moisture between grains of earth. *C. elegans* was the first animal ever to have its genome fully sequenced, in 1998, some two years before the fly and five years before us. It had been proposed as a model organism by South African biologist Sydney Brenner in 1963 to investigate animal development because, vitally, this tiny worm is one of the simplest animals to possess a nervous system. By the 1970s, Brenner had left South Africa and pulled together an impressive team of scientists to study and develop *C. elegans* as a model organism in odd corners of laboratories at Cambridge University in the UK.

'The history of the worm is full of the devouring lust for knowledge,' writes Andrew Brown, in his marvellous biography of this little creature, *In the Beginning Was the Worm*. 'The people who [studied] it were not really interested in money, or fame outside a limited circle. They were not saints. They were ambitious and competitive, and life was hard for those who failed. But their ambition and competitiveness and their sometimes jealous love were all directed at altruistic ends. They wanted to understand the world. They wanted measurable, solid truths about it.'

They may not have been seeking it, but fame these scientists achieved nevertheless. In 2002, Brenner and two of his worm colleagues, John Sulston and Robert Horvitz, shared the Nobel Prize for Medicine for their work on organ development and what's known as 'programmed cell death', as revealed by *C. elegans*.

The worm is a wonder of simplicity, having no brain to speak of, and no heart, though the muscular pharynx pumps like a heart. The hermaphrodite version has fewer than 1,000 cells, including 302 neurons. It is transparent, so the development of its body from the egg to adulthood can be watched as it happens. Unlike the fly, the worm can be frozen and thawed back to life, so that experiments can be interrupted and worms easily stored.

But its ability to withstand deep freezing is just one facet of the worm's extraordinary resilience. As one of the species sent into space for various experimental purposes, *C. elegans* survived the disintegration of Columbia in 2003 that killed seven astronauts when the shuttle returned to Earth. In 2009, this tiny tough scrap of life was sent out again, to the lab on the international space station, by scientists at Nottingham University, UK, keen to study the effects of zero gravity on muscle development. They were looking especially for insights into the genetic basis of muscle atrophy – issues of direct relevance to the bedridden, people with diabetes, and the elderly, for example.

The sequencing of its genome in 1998 revealed that the worm has around 19,000 genes arranged on six chromosomes. One can only imagine the scientists' surprise when they discovered, over the next few years, that this is larger than the genome of the much more complex fruit fly with its 13,600 or so genes, and not much smaller than our own genome, estimated to have between 20,000 and 24,000 genes. More than one-third of *C. elegans* genes have counterparts in humans.

In 1986, Sydney Brenner's lab at Cambridge announced that they had created a wiring map of the worm's nervous system. Taking their cue from the genome and its sequencing, they called this map a 'connectome'. It was created from photographs of electron-microscope images of wafer-thin serial sections of the creature's super-simple brain, and shows how every one of its 302 neurons is connected to others. Mapping the *C. elegans* nervous system marked the beginning of 'connectomics' as a discipline, the ultimate goal of which is to draw up a map of our own neural circuitry for research of all sorts into how the brain functions and how it begins to fail with age.

Before judgements are made about the relevance to human biology of discoveries made in these lower organisms – and there are many other 'simple' models besides flies and worms, including yeast, bacteria, viruses and several species of little fish – they need to be tested in mammals. Of these, mice have been around the longest and are the most widely used model organism in biomedical research today. In the 17th century the English doctor William Harvey used them to work out how blood circulates round the body, driven by the pumping action of the heart. They were used to study respiration in the eighteenth century. And there's a lovely story of how Gregor Mendel originally began studying the laws of inheritance using mice with different-colored coats which he bred in cages in his monastery cell. He switched to pea plants after receiving complaints from his buttoned-up superior about the 'smelly creatures that, in addition, copulated and had sex'.

Mice are stars in ageing research, too. But, according to evolutionary biologist Steven Austad, they share with flies, worms and most of the animals gerontologists use in the lab a central weakness: they're all short-lived. 'That's why they're useful to researchers,' he comments. 'But we could be missing some huge things by only looking at short-lived animals. And particularly, since we're long-lived, it may be that our biology has already taken advantage of all the things that people are discovering in worms and flies and mice. So I thought, why don't we try to look at something that ages more successfully than we do?'

Austad, now in his sixties, is slightly built, with thinning ginger hair and a small moustache. He has the air of someone with insatiable curiosity, always on the lookout for adventure and new ideas. It's not surprising, therefore, to learn that his path towards a career in science was anything but conventional, and involved a goodly dollop of impossible dreams and risk-taking. 'I wanted to write the great American novel,' he told me with a grin, as we sat talking over cups of coffee on the fringe of a gerontology meeting in New York. So he took a degree in English, and then a series of odd jobs while he worked on the novel-that-was-never-to-be.

'One of the jobs I ended up doing was training lions for the movie business.' Enjoying the startled look on my face, he continued, 'What happened was … I was working as a journalist in Portland, Oregon, and I had a friend who had two pet African lions. He got an offer to use them in a Hollywood movie and he needed someone to help him transport them. So we took the back seat out of his car, we put a lion in there and we drove a thousand miles [1,600km] to Hollywood. When we got down there the producer offered me a job. I said, "I don't know anything about this", and he says, "Well, I've hired experienced trainers and you could learn from them" … So I took that job.'

Austad stayed three and a half years in Hollywood, eventually working with 56 lions – sometimes as many as 25 animals together in one movie. 'We had to work on group dynamics,' he said, 'and doing that sort of awakened an interest in science that I'd had already, because I'd started off as a math major. So I decided to go back to school and get a PhD.' He had time to make and mull over that decision during a long stay in hospital after one of his lions turned on him during a training session, when he briefly let slip his psychological dominance over the big cat. It severely mauled his leg before Austad, in a tangle with the lion, was spotted by one of his colleagues and the animal was driven off with a noisy fire extinguisher. 'The doctor told me I'd never walk normally again. But he was wrong about that, I'm very pleased to say.'

Austad's doctoral thesis was on animal behaviour and ecology, but his plan to do field studies with lions in Tanzania's Serengeti National Park didn't work out, and he ended up studying opossums in Venezuela instead. It was here that he became interested in ageing as a phenomenon. 'In the course of the project I would recapture these animals every month and I discovered that they age incredibly rapidly,' he explained. 'They age as fast as a mouse. I didn't know at that point how fast a mouse ages, but I knew these things aged a lot faster than you'd expect. They'd get cataracts and parasites and they'd lose muscle mass – and I could see this in a few months.'

Ageing research, Austad figured, when the time came to move on in his career, had a good future, 'since anyone could see what was happening to the demography of the world'. His first ageing study involved field work with wild mice. But curious about what we might be missing in using short-lived organisms as models for ageing in us, he changed his focus. Today he studies what is believed to be the world's longest-lived multicellular animal, *Arctica islandica*, a species of cold-water clam (also known as an ocean quahog) found in the muddy bed of the North Atlantic Ocean all around the British and Irish coasts, in the Baltic and off the US coast from Cape Cod northwards.

It all started with a phone call from a bunch of marine biologists in the UK who had heard of his interest in ageing research and asked if he'd like to collaborate with them on their studies of long-lived clams, Austad told me in New York. 'I said, "What d'you mean by long-lived?" They said, "400 years." It just blew my mind,' he laughed. 'In fact I think I said, "I'm sorry, I thought you said 400 years!" And they said, "We did."' The marine group were studying ancient climates and Austad leapt at the chance to work with their ancient clams too.

As with trees, the age of a clam is calculated by counting the growth rings in its shell. Most of those dredged up from the ocean by his new colleagues were 100 to 200 years old, with just a few found to be older – up to 400 years. This was thought to be about the limit, until Ming the Mollusc came on the scene in 2006. Originally thought to be around 405 years old, the clam – named Ming after the Chinese dynasty in power at the time of its birth – hit the news headlines and gained a place in the Guinness World Records. But when a more sophisticated counting technique became available in 2013, Ming was found to be more than a century older than first reported – at fully 507 years. The mollusc might well still be alive today but for the fact that you can't tell the age of a clam accurately until it's dead and you can remove its shell to study the hinge area. So Ming had been frozen to death on board the research ship by its British discoverers.

But what have these ancient molluscs been able to teach us about the secrets of an exceptionally long life? Researchers have looked at all kinds of physiological processes, including respiration and metabolism, for clues. What stands out most clearly as special in the molluscs – and indeed in several other long-lived species that have been studied – is the extraordinary stability of their proteins, the products of genes that do practically all the work in an animal's body.

'Proteins, to do their job, have to be intricately folded, *precisely* folded like origami,' explains Austad. 'And what happens is that, over time, your proteins in some cells – particularly in your cells that last a long time – gradually get misfolded and degraded and that can cause them to become toxic as well as dysfunctional.'

Misfolded proteins can stick together to form clumps that our bodies find hard to eliminate. Alzheimer's disease is a classic example: here the protein beta-amyloid aggregates to form sticky plaques in the brain that are associated with the death of nerve cells – a scenario that will be described in a later chapter. Ironically, the ocean quahog, which doesn't have a brain, may suggest ways of preventing or clearing the sickly proteins of dementia. Interested to see whether the clams could stabilise the proteins from other organisms too, researchers bathed some beta-amyloid in clam 'juice' – the extracted contents of muscle cells – and found that the amyloid was unable to clump together. 'They've got some mechanism for stabilising all proteins – even proteins from people,' commented Austad. 'So that's why we think they may turn out to have therapeutic value for things like Alzheimer's disease.'

So far, the scientists have been unable to pin down the magic ingredient in the clam juice. To make real progress, says Austad, they need to be able to analyse the animal's genes, and as yet they have only a rudimentary first draft of its genome to work with. What scientists will look for, once they have a reasonable road map, is genes that we share with the clam, but that seem to be working more effectively in them than in us. Such a strategy has, very recently, answered a question that has long intrigued scientists: why, with so many more dividing cells in their huge bodies, elephants don't have higher rates of cancer than we do. It turns out that elephants have 20 copies of p53, the tumour-suppressor gene we met earlier, whose job is to eliminate cells that get damaged during the process of division (humans and other mammals have just one copy of the gene). What's more, says Austad, elephant p53 is a more sensitive version of the gene than ours. 'So if you damage DNA just a little bit, elephant cells are more likely to start the suicide programme than human cells are. That's the kind of thing we're looking at for these protein-stability molecules in *Arctica islandica*.'

Gerontologists are studying DNA for clues of all sorts to why and how we age, and what we might do to avoid the problems and pains of our degenerating bodies – and they're reaping some rich rewards. So who are the researchers and what are they finding?

CHAPTER SEVEN

It's in the genes

In 1993 *C. elegans* hit the headlines and gave a shot in the arm to ageing research when Cynthia Kenyon, now Professor of Biochemistry at the University of California, San Francisco, discovered that a mutation in a single gene, known as daf-2, could double the lifespan of the little worm. Here was evidence that the ageing process is under some sort of control and is not simply the random wearing away of our bodies by time. Tinkering around with daf-2 over the following years, Kenyon's lab managed in 2003 to extend the little worm's lifespan by sixfold, and later another lab was able to extend it tenfold. 'You look at these worms and think, "Oh my God, these worms should be dead." But they're not,' she told science writer Bill O'Neill in 2004. 'They're moving around ... Once you get your brain wrapped around that ... then you start thinking, oh my goodness, so lifespan is something you can change – it's plastic. Then who knows what the limit is?'

Dramatic discoveries in science are almost invariably the result of long, hard slog at the lab bench by many people, and Kenyon was standing on the shoulders of giants. Michael Klass was the first person, in 1983, to write about variable lifespans in different gene mutants of *C. elegans*. But he reckoned it was the indirect effect of the mutations on the worms' eating habits that made the difference. Then in 1988, Tom Johnson and David Friedman at the University of Colorado, Boulder, pinned down the effect to the function of mutant genes themselves, and identified one in particular, which they named age-1, that gave carriers of the mutant version an extra 40–65 per cent of time on Earth. This was the first ever gene mutant to be identified in any living species that produces a longer lifespan. Longevity in Johnson and Friedman's worms, however, came at the price of poor sexual function and impaired fertility. Kenyon's daf-2 mutants, on the other hand, simply aged more slowly while remaining as healthy, vigorous and fertile as non-mutants.

Turning to a video clip she'd put up on the big screen at a TED talk about her work in 2011, Kenyon told her audience, 'In just two weeks, the normal worms are old. You can see the little head moving down at the bottom there. But everything else is just lying there. The animal's clearly in the nursing home. And if you look at the tissues of the animal, they're starting to deteriorate ...' She clicked through images to another video clip, 'Now here is the daf-2 mutant,' she said, adding, to a peal of laughter from her audience, 'one gene is changed out of 20,000. It's the same age [as the normal worms], but it's not in the nursing home. It's going skiing.'

This is a huge milestone in the field of gerontology, so let's follow it from the beginning. When Michael Klass, Tom Johnson and their colleagues began looking in the 1980s for genes that might be involved in ageing, they were right out on a limb scientifically. 'At the time,' writes Cynthia Kenyon in a personal review of the discovery of the long-lived mutants, 'ageing was thought to be a hopelessly intractable, even futile, problem to study. We just wear out; that's it.'

Those who did study it were mostly dismissive of the idea of searching for clues in the DNA, since ageing occurs after reproduction when the forces of natural selection have run out of steam – implying that the process of decay has nothing to do with the genes, which are the programme for living, not dying. But Johnson's discovery had blown a hole in this assumption and Kenyon was excited. 'I saw the analysis of ageing as a fantastic opportunity to explore the unknown and perhaps discover something new and important,' she wrote.

At genetics and worm meetings around California, she had heard Tom Johnson talk about his age-1 mutants and had been particularly intrigued by his assumption that the lifespan effect was linked to their impaired fertility. This was what the disposable soma theory would have predicted: that, in the resource budget of living, the mutant creatures had more to spend on looking after their bodies because of the unusually low demand from reproduction, and therefore they lived longer. But could it really be that simple, that direct an equation?

Kenyon set out to investigate. She assigned one of the grad students in her lab, Ramon Tabtiang, the task of zapping with a laser-equipped microscope the latent egg cells in newly hatched *C. elegans* (thus suppressing reproduction) to see what it did to their lifespan ... Which was nothing. Clearly there was no direct balance to be struck here between reproduction and lifespan. But what *was* going on? What mechanism was being triggered or suppressed by the mutant gene to give the worms a longer life? She was determined to find out.

Kenyon had first become interested in ageing as a phenomenon when she was working as a postdoc in the lab of Sydney Brenner – the 'father of worm biology', you'll recall from the last chapter – in Cambridge, UK, in the early 1980s. Leaving a dish of worms in the incubator one day while she got on with other work, she was surprised to note, on retrieving them from the incubator several days later, that the tiny creatures looked decidedly old. 'This concept, that worms get old, really struck me,' she wrote. 'I sat there, feeling a little sorry for them, and then wondered whether there were genes that controlled ageing and how one might find them.'

When she began her research later in California, ageing was still considered 'a backwater by many molecular biologists'. One of her colleagues even warned her she would 'fall off the edge of the Earth' if she pursued her interest. But she and Tabtiang took no notice, and they were amazingly lucky. Unwilling to trespass on someone else's territory, as soon as they had checked out the reproduction–lifespan conundrum with Johnson and Friedman's age-1 gene, they started looking for long-lived mutants of their own to work with. Painstakingly screening worms with promising characteristics, they managed within a surprisingly short time to isolate one strain that outlived all the others and that had the mutation in a single gene they later named daf-2. 'daf-2 mutants were the most amazing things I had ever seen,' writes Kenyon in her review. 'They were active and healthy and they lived more than twice as long as normal. It seemed magical but also a little creepy: they should have been dead, but there they were, moving around.'

So how does the gene work? It turns out it's part of what's called a 'nutrient-sensing' network, which does as its name implies: it monitors the organism to see that it's got enough nutrients on board to fuel its vital activities. The daf-2 gene is the recipe for a structure on the surface of cells (a so-called receptor molecule) that is the entry point into the cell of important hormones – insulin and growth factors. The insulin enables tissues to absorb nutrients and turn sugar into energy, and the growth factors promote the building of proteins for growth, maintenance and repair, and many other tasks.

Mutant daf-2 builds faulty receptors that limit the amount of hormone getting into the cell. This sets off an alarm system suggesting the available fuel doesn't meet the creature's needs, and it must take defensive action. The alarm signal activates a caretaker gene called FOXO, which wakens from its usual slumber inside the cell, gets dragged into the nucleus where the DNA is tightly packed, and switches on a bunch of other genes. The tasks of these genes are, variously, to protect the cells from oxidative stress (free radicals); to repair or recycle damaged components; and to make sure that other proteins in the cell are formed and functioning properly. This scenario made intuitive sense to researchers, because already the gradual failure of maintenance and repair of our bodies had been identified as one of the defining characteristic of ageing – if not the whole story. But the big question was, how representative were these findings in the worm? Did they offer real insights into what might be happening in other creatures – including us?

'Biology is very reductionist,' says David Gems, whom we met back in Chapter 2 taking potshots at the oxidative damage theory. 'I mean, gene function – how genes work – wasn't solved using human genes; it was solved using bacteria, and bacterial viruses even, to work out things originally. The idea is that you start with something simple and work up; if you can figure that simple organism out, it should be a big starting point to branch off. I still think that's correct. *C. elegans* is an animal with a nervous system, embryogenesis and sexual reproduction, muscles and all of those things, and I think it's been amazing in terms of discovering fundamentals about ageing.'

Gems was drawn to ageing research as a biology student in the early 1980s at the University of Sussex, where he was much influenced by the writings of Thomas Kuhn, the American physicist and philosopher. Kuhn spoke of revolutionary science driven forward by imagination and conceptual leaps – or 'paradigm shifts' – rather than by slow, steady incremental steps of knowledge. Gems wanted to go where the revolutions were most likely to happen, where the big unanswered questions were. After graduating, he pushed off to see the world, spending two years travelling, working and tangling in political revolution in Latin America. He returned to the UK in the late 1980s to do a PhD in genetics at Glasgow University, before finally turning to gerontology and taking up with the little worm at the University of Missouri in the US in the mid-1990s.

When it came to choosing a career in science, Gems wanted to do something so compelling that it would drive him throughout his life, keep him excited. Ageing fitted the bill. 'The main cause of disease in the world! Why would that not be the most important question in biology, practically?' he says. 'I mean, maybe consciousness is bigger. But medically there is nothing greater, and it's bizarre that it's been neglected. I realised, my God, this is a huge topic that is completely neglected! This is amazing.'

Since 1997, when he moved back to the UK and to University College London, Gems has been deeply occupied with identifying the cat's cradle of genes – thousands of them, as it turns out – regulated by the caretaker gene FOXO as part of the nutrient-sensing network in *C. elegans*, and trying to discover how exactly this system can bring about the changes in cells, tissues and bodies that we see as ageing. 'The idea is that you find the genes that control ageing and then you discover what ageing is. But it's turned out to be very difficult,' he laughs, pushing towards me a piece of paper scribbled with lines like a mighty telephone exchange that represents the interactions he is looking to understand.

Working next door to David Gems is Linda Partridge, Director of the Institute of Healthy Ageing at UCL. While Gems's focus in the search for ageing-associated genes is the tiny worm, Partridge's model organism of choice is the fruit fly *Drosophila.* A thoughtful, soft-spoken woman in her sixties, who wears colourful owlish specs, Partridge became fascinated with science during her schooldays. 'Finding out how the world works; how people set about trying to force Nature to give up its secrets …' she mused as we sat talking in her office in London, 'I must have read a few, not whole biographies by that point, but accounts of scientists' lives and how they'd stumbled on things, and I found the whole process, the kind of determination and ingenuity they had to show … I was just full of admiration.'

Growing up on the cusp of social change, when marriage and motherhood were still the destiny, if not the dream, of so many young girls, Partridge says she never imagined she would become a scientist herself. But she was bright, had a supportive and open-minded family, and went to Oxford University. Today she is an original mind and a widely respected figure in geroscience who divides her time between UCL and the Max Planck Institute in Cologne, Germany where she is a founder member of the ageing research institute.

Partridge is an evolutionary geneticist by training, and says, 'Ageing is weird from the evolutionary point of view, because what you have is an obviously maladaptive trait. You've got an organism that's been designed to develop in a very orderly fashion to give rise to an adult, and then it falls apart. You'd think it an awful lot easier to keep an organism working than to build it in the first place. So *why* do we get this process of decline and death?'

As the role of the nutrient-sensing network in *C. elegans* was being uncovered in the late 1990s, Linda Partridge and David Gems mused as to whether the same network in her model organism, *Drosophila*, might also have a connection with lifespan and ageing. Partridge thought it a long shot. But she and Gems got the opportunity to test the question when another colleague, Sally Leevers, interested in mechanisms of cell growth and proliferation in relation to cancer, was investigating the nutrient-sensing network in flies and discovered – quite by chance – a single gene that had a profound effect on growth: when mutated, the gene produced dwarf flies.

Because of the light it shone on the internal workings of cancer cells, Leevers's discovery 'led to a feeding frenzy [among scientists] and the isolation of other genes in the pathway,' said Partridge. But no one was watching to see if their mutant flies lived longer than usual. 'I don't think they were in the least bit interested – that was considered a quaint little corner of science!' she laughed. People were, however, happy to share their mutant fly strains with the UCL researchers, who soon identified a gene that extended the lives of the flies by up to 48 per cent when mutated or knocked out altogether. They named their new gene chico because of its effect in producing diminutive flies, and published their results in *Science* in 2001. Like daf-2, chico works high up in the nutrient-sensing network and it triggers the same cascade of events, awakening the caretaker FOXO to go to the nucleus and switch on a battery of protective genes.

'One of the obvious conclusions from the evolutionary work,' says Partridge, 'is that you'd expect a lot of different genes to contribute to the ageing process. So when these single-gene mutants suddenly appeared that could extend lifespan, I was very intrigued. We hadn't really been expecting those.' Almost no one else had either, so the discovery of age-1, daf-2 and then chico – all of them working through the same core signalling network in the different creatures – caused something of a revolution by shattering many assumptions. It begged the obvious question: how far did this phenomenon found in the invertebrates extend through the hierarchy of species?

Everyone wanted to know, and within two years a report had come in of mice whose lifespan was extended by 18 per cent by knocking out altogether a single gene that holds the recipe for an insulin receptor specific to fat cells. Soon thereafter, mutations in a number of other single genes involved in the nutrient-sensing network were found to give mice extra months of life. But much more important than extra days on Earth was the finding that 'most of the things that go wrong in ageing go wrong more slowly in the mutants,' said Partridge, speaking at a Molecular Frontiers symposium in Sweden in 2016.

To illustrate her point, the scientist threw up a slide on the big screen showing two mice, littermates, only one of which had been genetically modified to knock out the insulin receptor. The mice were aged a little over two years, and whereas the normal mouse looked decidedly old, with rumpled fur, the beginning of cataracts, the hunchback of osteoporosis and rickety legs, the other looked youthful and sprightly. The protection from disease in such a broad spectrum of tissue and bodily systems that seem to have no connection to one another is particularly intriguing, commented Partridge. It implies 'that we really have hit something about the underlying ageing process'.

But what about humans? Does any of this have direct relevance to us? The obvious way to find out was to focus on the nutrient-sensing network and test the genes in us that are analogous to those shown to influence lifespan in the model organisms. Whereas invertebrates have only one of the caretaker FOXO gene, humans and other mammals have four. And sure enough, in 2008, a group of scientists led by Bradley Willcox at a research institute in Honolulu, Hawaii, published a paper showing a strong association between a natural variant of the FOXO3 gene and long life. Their study population was a group of 213 men, Americans of Japanese origin, over the age of 95 years.

These men they compared with 402 controls from the same community, none of whom survived beyond 81 years during the study period. The variant, FOXO3A, stood out as the most significant difference between the two groups in a bunch of five candidate genes. And it looked like good news on many fronts for those who had inherited the gene: they had less cancer, cardiovascular disease and cognitive decline than the controls, and were physically stronger and steadier on their feet, despite being on average 11 years older.

Since that first paper from Willcox and colleagues, evidence of the beneficial effects of FOXO3A on health and lifespan has come from population studies among Han Chinese, Ashkenazi Jews, Californians, Germans, Italians and Danes. Researchers studying centenarians among Ashkenazi Jews in New York City have also found a link between their extreme longevity and mutations in a specific cell-surface receptor for insulin and growth hormone. Sounds familiar? This is indeed the human version of what Kenyon and colleagues found in their worms with daf-2 mutations.

It makes sense that if you live to an extreme old age you must have managed to stay healthy for longer. But the empirical evidence of the relationship between healthy ageing and longevity is striking and thought-provoking. A study by a group in Boston looked at the medical histories of nearly 1,500 centenarians – 534 of them considered the oldest old, aged between 105 and 119 years. They compared the centenarians with 343 people aged between 97 and 99 years, and 436 controls aged 47–96 years, and found that the older a person was at death, the later in life they were likely to have suffered their first serious age-related disease – defined as cancer, cardiovascular disease, chronic lung problems, diabetes, dementia or stroke – and the shorter the period of their total lives they had spent as frail old people.

Thus, on average, the proportion of their lives spent with serious incapacity was 9.4 per cent for those who died in their late nineties; 9 per cent for those who died between the ages of 100 and 104 years; 8.9 per cent among those who died aged 105–109; and only 5.2 per cent for the oldest old, those who died between the ages of 110 and 119. In fact, 10 of the 104 supercentenarians escaped serious disease right up until the last three months of their lives. By contrast, the controls whose lifespans were not considered exceptional suffered chronic ill health on average for 17.9 per cent of their lives.

But back to the genetics, and at this point I need to strike a note of caution. There is no suggestion that what's been uncovered regarding nutrient-sensing genes – a pattern that seems to hold true from the humblest organism on Earth to the most sophisticated – is any sort of 'key' to ageing, any sort of master switch. There is abundant evidence that no such master switch exists. Genes such as age-1, daf-2, chico, the FOXOs and their equivalents in other critters form the tip of the iceberg. The search goes on for the multiple driving factors that lie beneath. As of mid-2017, scientists had identified 2,152 genes that affect lifespan (and therefore the rate of ageing) in model organisms, and 307 such genes in humans. From what they have learnt so far, they reckon that some 20–30 per cent of our natural lifespan is accounted for by our genetic heritage, while environmental influences account for the rest. Interestingly, a number of the lifespan-influencing genes identified in both model organisms and humans involve another key player in the nutrient-sensing network – the TOR signalling system that we first met in Chapter 4 in relation to senescent cells.

The compelling message to be drawn from all this work on genetics, David Gems told an interview for *The Naked Scientists*, is that 'ageing is plastic. It's not fixed. It's something that is alterable.' And that discovery is 'very, very profound. It has shaped a whole lot of thinking in the field.' And it has given tantalising clues to what might be possible in minimising the miseries of old age.

All kinds of new ideas have bubbled to the surface, including one that suggests that the vexed theory of oxidative damage – which sees the by-products of metabolism, the free radicals generated by burning sugars in the cells as fuel – was always too narrow in its focus. After all, the argument goes, the regular metabolic processes that generate free radicals are what keep the engines of every living creature ticking over; ways of dealing with the inevitable by-products of burning sugars for energy are built into the system, like fuel filters in a car engine.

There are, however, many other random and unpredictable threats to the engine that must be dealt with as and when they occur, with whatever tools the body can find. These threats come from the environment in which we live, and include, for instance, the food we eat and the chemicals with which we come into contact. Neutralising these external (or 'xenobiotic') threats to the cells is far more of a challenge to our bodies than free radicals. Inevitably, we'll be less than successful in counteracting such threats some of the time, and damage will accumulate over the years.

This argument is compelling, and today those scientists still somewhat hooked on the 'wear and tear' paradigm of ageing are looking at a far broader spectrum of damaging processes. But Gems and Partridge have studied the genetic data and proposed an altogether more radical – indeed revolutionary – interpretation. What if, instead of endorsing the theory that ageing is caused by accumulation of damage and failure of maintenance systems, the data are pointing to the exact opposite – that ageing is driven not by *failure* of systems but by their overexpression? In other words, by natural processes of living that simply run on too long?

This is the 'hyperfunction theory' of ageing, and it was first suggested in 2008 by biologist and cancer specialist Mikhail Blagosklonny. At its heart is the idea that ageing is driven by the action of normal genes (known in the business as 'wild-type' genes), not mutants – the same genes that drive our development from egg to adult, through reproduction. These genes continue to work beyond the development programme for which they were selected, but in the post-reproduction period of life they are under ever-weakening evolutionary control and their no-longer-appropriate activity leads in time to disease and death.

'What the biology is saying,' explains Gems, 'is that it's your wild-type genes that are causing pathologies. They bring you into existence; they take you to maturity and then they carry on regardless and they generate pathologies.' Looked at from this viewpoint, ageing *itself* is a disease process, and the clear pathologies we recognise as age-related diseases – the cancers, heart, vascular and brain problems and all the rest – are not so much separate entities, deviations from the norm, as just the most distressing points on a continuum of disease.

A fascinating idea indeed. So what is the evidence for the hyperfunction theory? Gems and Partridge point out that many of the classic pathologies of old age involve runaway growth or over enlargement of cells rather than decline and decay. This is the case with cancer, cardiovascular disease, diabetes and Alzheimer's. Excess growth is a common cause also of the prostate problems that plague many older men as the tube leading from the bladder gets squeezed. And even the thinning bones of osteoporosis can be caused by hyperactivity of the cells responsible for breaking down bone in the process of sculpting our skeleton, the osteoclasts, which get out of kilter with the bone-forming cells, the osteoblasts.

Our little worm, *C. elegans*, too provides evidence of hyperfunction. Being a hermaphrodite, it produces both sperm and eggs, switching from sperm production to egg production in early adulthood in preparation for self-fertilisation. Reproduction ceases when the finite sperm supply has been exhausted. But the tap that supplies the egg cells with nourishing yolk continues to run, accumulating eventually to toxic levels. 'The worm fills up with these oily pools of yolk, which is a kind of obesity,' says Gems. 'We know that the wild-type genes control yolk production, so you can say: well, they're just continuing to do that beyond what is useful, hence you end up with these pools of yolk. It behaves as if the tap just keeps running. There's no point at which anything has gone wrong; it just continues to do what it's supposed to do.

'In the worms these pathologies of ageing develop very, very quickly, because it's a short-lived animal, so it's a little bit simpler. In higher animals it's less clear … I mean, it could be that the tap is pouring [in the worm] and everything goes very quickly. But in a higher animal it's more of a drip, drip, drip, and gradually there's a cumulative change; you end up with a cascade of things triggering further change.' Significantly, worms with daf-2 mutations don't fill up with yolk, and live extra-long lives.

So where does the hyperfunction theory leave the issue of wear and tear? David Gems believes molecular damage *does* play a part in ageing, but generally as a consequence, or perhaps even a trigger, of the process rather than a driving force. He draws the analogy of a hand grenade that has blown a room to bits. Molecular damage is the grenade pin, and the wild-type gene activity is the TNT. 'If you ask what caused the damage – was it the pulling of the pin, or the TNT? – you'd say, well actually the TNT was the main thing,' he comments. Cancer is a particularly good example. Damage to the DNA is what kicks off the process, but it's the rampant growth of cells run amok that forms the tumours that kill.

Linda Partridge frames the same argument in terms of antagonistic pleiotropy (remember that from Chapter 1? To recap, this is when a gene that has beneficial effects in young, developing organisms has harmful effects in later life). 'The nutrient signalling network is very important for growth and reproduction, which are young traits,' she explains. 'But it seems to be set at a level that drives cells too hard once they [are no longer young]. Maybe they've got a bit of DNA damage; maybe the regulation of proteins is starting to break down, and then suddenly being hit with this signalling pathway which says "Activate everything; make stuff" … The cell can't handle it and it causes more damage than it would in a functional, young cell.'

Hyperfunction theory is not yet mainstream and most genetic research into ageing is still focussed on the nutrient-sensing network. One way of manipulating this that has fascinated scientists for some time – and found favour with a smallish band of determined 'immortalists' around the world – is to cut back more or less dramatically on the amount of food eaten. This strategy is known as calorie, or dietary, restriction, and it has a colourful history.

CHAPTER EIGHT

Eat less, live longer?

Way back in 1935, scientists at Cornell University, Ithaca, New York, discovered a crucial piece of the great jigsaw puzzle of ageing, though its important position in the developing picture would take another half century to be widely recognised. Clive McCay, an American biochemist and nutritionist, and his colleagues were interested in the relationship between rate of growth, stature and lifespan, and were busy investigating this relationship in a bunch of lab rats.

The scientists already knew that rats given a diet lacking essential nutrients not only grew more slowly than normal and were stunted as adults, but that they tended to get sick and die early. But what would happen, they wondered, if you gave the rats all the nutrients they needed to develop to maturity, but simply cut back on the calories to slow the process down? Would the animals eventually grow to the same size as those allowed to eat as much as they wanted? And would the rate of growth, whatever size they were at maturity, have any effect on their lifespan?

Working over a four-year period with 72 rodents assigned to two different levels of calorie restriction (36 creatures to each group) plus a group of 34 controls given a full-calorie diet and allowed to feed at will, the scientists found dramatic differences in lifespan. Some of those on the most restricted diet lived more than twice as long as the controls. (Generally speaking, this effect was much more marked in the male rats than in the females – a point I'll return to a bit later).

McCay was Professor of Animal Husbandry at Cornell and the primary focus of his lab's research was improved production among America's beef and dairy herds. But McCay himself had long been interested in the biology of ageing and he clearly saw that, besides its usefulness to the cattle industry, what he had discovered in his lab rats was a brilliant new tool for studying his other obsession. 'Retardation of growth by diets, complete except for calories, affords a means of producing very old animals for studying ageing,' he wrote rather soberly in a paper in 1939.

In a 2010 review of 75 years of calorie-restriction research, Roger McDonald and Jon Ramsey of the School of Veterinary Medicine at the University of California, Davis, credited the strategy, somewhat contentiously, with contributing 'more than any other model to the overall understanding of the biological processes of ageing and longevity', and even described it as 'one of the greatest discoveries ever made in biology or medicine'. Be that as it may, the idea of calorie restriction took a mighty long time to be recognised: it wasn't until some 40 years after McCay's paper appeared in *The Journal of Nutrition* that anyone else saw the relevance of his rat work for ageing and began studying it for themselves. Among the new enthusiasts was Roy Walford, the eccentric Professor of Pathology at the University of California, Los Angeles (UCLA), who was to become the public face of calorie restriction.

Walford originally used the strategy to produce a batch of healthy old mice with which to do research on the immune system. His lab was one among several to show in the 1970s and '80s that not only does calorie restriction (or CR as it is better known to aficionados) extend the lifespan of rodents, but it also delays – often dramatically – the onset of age-related frailty and other pathologies, including cardiovascular diseases, cancer, diabetes and neurodegeneration. In other words, it gives rats and mice extra years of vigour and disease-free life. (So much so, in fact, that researchers today say you often can't tell what a calorie-restricted mouse has died of – it just seems to run out of steam: one day it's fine; the next it's dead, without any obvious pathology).

The assumption at the time Walford began investigating CR was that for the regimen to extend lifespan it had to be started in young animals, and indeed researchers who tried dramatically cutting the calories of adult mice ended up with creatures that soon got sick and died. He and his grad student Rick Weindruch, however, tried cutting the calories of adult mice gradually over a three-month period and managed to extend their lives by 20 per cent. This was a clincher for Walford: what worked for rats and mice (and apparently water fleas too!) would probably work in humans, he reckoned. In 1984 he adopted a CR diet himself. The recommended daily calorie intake for men is around 2,500 a day; Walford restricted his to around 1,600, carefully balanced between nutrients to ensure he suffered no deficiencies.

He wanted to live to at least 120, he would tell reporters fascinated by this colourful scientist who shaved his head, sported a droopy moustache, rode a motorbike and had a habit of dropping out periodically from his busy lab work for bouts of adventure.

Walford travelled the African continent on foot, and once walked across India in a loin cloth – 'measuring the rectal temperature of holy men', according to his many obituary writers who couldn't resist this bizarre detail. Expounding his 'Signpost Theory of Life', Walford told the *Los Angeles Times* that years spent simply grafting at the lab bench, even if it might lead to a Nobel Prize, would pass in a blur. So he found it 'useful to punctuate time with dangerous and eccentric activities'.

Of these activities, the one that attracted most publicity was his time as team physician among a group of eight people – four men and four women – who holed themselves up for two years in a human terrarium in the scorching

Arizona desert in 1991, in the name of science. This was Biosphere 2, a prototype space station intended to test the feasibility of living for extended periods on other planets. Biosphere 2 (our Earth is Biosphere 1) was a sealed ecosystem, 1.27ha (3.15 acres), under a glass dome that contained five natural biomes – representing rainforest, savanna, desert, ocean and bog – as well as an agricultural station and living quarters. The technological paraphernalia needed to run the operation was housed in a basement.

Biosphere 2 was supposed to be self-sustaining, with air, water and organic material being recycled, and food produced on its own farm. This was a tall order, however, and over the two years the crew of the pseudo space station lost more than 20 per cent of the 3,800 animal species (including insects) they took in with them, and they ran distressingly low on oxygen – down to 14.2 per cent of normal levels. But of most relevance to our discussion here, they also struggled to produce their food, and Walford, responsible for their health and well-being, had to impose a heavily restricted diet. He was able to work out a regimen that gave them all the nutrients they required to keep going, but less than 1,800 calories per person per day.

By all accounts it was a stressful existence of constant, nagging hunger exacerbating other emotional tensions between the inhabitants of Biosphere 2. Everyone lost a lot of weight, but what excited Walford was that they showed some of the important physiological improvements that had been shown in rodent studies with calorie restriction, including dramatic lowering of blood pressure and cholesterol levels, and more efficient processing of glucose. And they didn't get sick, despite the fact that CR has been found subsequently to dampen the immune system, making people less able to fight off viruses or heal wounds.

Walford's belief in CR as the route to a long life and healthy old age was strengthened by the data from

Biosphere 2. Hunger, he felt, was a small price to pay for what he, personally, experienced as a wealth of benefits. He felt healthier on the diet he'd pursued for a decade and more, he told the BBC's Peter Bowes, who visited him at his home in Santa Monica, California, some years later. He needed less sleep, felt intellectually stimulated and full of a sense of well-being and vitality. 'If you want to trade all that to eat cake, then I say go ahead and eat cake,' he commented.

But Roy Walford never did reap the rewards of his monumental self-control. He died in 2004 of motor neurone disease, aged 79. (Ever a believer, he reckoned CR had slowed the progress of his fatal illness and given him extra months). He did, however, help to put CR firmly on the ageing research agenda – and, with his prolific writings, to fire the imagination of millions of ordinary people who dream of much longer lives, if not immortality. Many have been inspired to adopt his spartan lifestyle. They call themselves CRONies (standing for calorie restriction with optimum nutrition) and today some 7,000 of them belong to the US-based international CR Society, which generates lots of useful data for researchers studying the effects of a spartan diet on human beings.

Dean Pomerleau has been a CRONie since 2000. 'When I started there was a growing and vibrant community of people on the CR Society mailing list, and it was a very exciting time,' he said when he spoke to me on Skype from his home in Philadelphia. 'There were high hopes CR could add 20 or 30 years to your life – extrapolating from some of the very early rodent models – and that was very appealing to a bunch of us. It was sort of a scientist's diet kind of thing … It was like we were exploring a new territory, applying some of this cutting-edge science in lower animals to humans in the hopes of a big pay-off down the road.' He broke into a grin. 'We were all very happy with delayed gratification – that was one of the prerequisites!'

Pomerleau was then an ambitious entrepreneur who ran his own technology company and was a key player in the effort to develop driverless cars. A computer scientist, he wrote his doctoral thesis at Carnegie Mellon University on how neural networks could process information from the road, a technology that has been used to develop anti-collision devices. In 1995 Pomerleau and his fellow researcher Todd Jochem took their rudimentary self-driving car, Navlab, on the road, travelling 4,500km (2,800 miles) from coast to coast on an expedition they dubbed 'No Hands Across America'. Pomerleau gets a real buzz from technology, and the future it promises is part of what motivated him to take up CR. 'I saw on the horizon a lot of exciting things and didn't want to miss out on them,' he laughs. 'I wanted to be around to see all of the cool science and technology I thought would be happening in 30 or 40 years' time, so I wanted to be healthy.'

At the time he began his dieting regimen, there was an idea in vogue – championed by the maverick, extravagantly bearded and whippet-thin gerontologist Aubrey de Grey, co-founder of the Methuselah Foundation – known as 'longevity escape velocity'. This basically suggested we could find ways to extend lifespan at a faster rate than time is passing, effectively outstripping death. 'So it wasn't just that calorie restriction might add 10 years; it might be that it would put you over that hump to live an extraordinarily long time!' said Pomerleau. 'So that at the time – and up until very recently – has been my motivation. And the motivation of many people doing CR.'

Today he eats only once a day, very early in the morning, a diet consisting of fruits, vegetables, nuts and seeds, including the occasional treat like a bulb of raw garlic, and no longer remembers what hunger feels like. 'I haven't been hungry in many years,' he says. 'I think the body has amazing resilience and ability to adapt to whatever regime you subject it to. I've been doing this for so long that hunger just decided to pack up and go home! It gave up trying.'

Dean Pomerleau, who is charming, outgoing and relaxed to speak to, and very ready to smile at his eccentricities, admits to being a bit of an obsessive. He is fascinated by the science, and after poring over the wealth of material from animal studies, he concluded that subjecting the body to cold has a synergistic effect with CR, encouraging the development of brown fat that helps burn calories. So he sets the temperature in his basement office at home about 10°F (5.5°C) below comfortable levels and works in short sleeves. He also periodically dons an ice-packed vest, taking his personal regimen to beat time beyond what's advocated by scientists in the lab.

What effect, I wondered, does his decidedly unconventional lifestyle have on his social and family life? 'I did find, very early, when I was just getting into calorie restriction, for the first couple of years I was a proselytiser. I was pretty fanatical about discussing it with anyone within earshot whether they wanted to or not!' he laughs. 'But I quickly learnt that that wasn't the way to make friends either within our family or with colleagues at work or friends outside of work, so I toned that down and it has been much easier since then.'

But back to the lab bench. Over the decades, CR has been investigated in a wealth of organisms besides rodents, including our old friends worms and flies, whose very short lifespan has enabled scientists to explore the mechanisms responsible for the effects of CR. (Experiments have also been done with rhesus macaque monkeys, intended to give a better indication than any of the other models of whether CR might indeed give us humans extra years of life as well as health. But more of them later).

This is a mighty complex task. First, taking account of all the variables in experiments with diet is a huge challenge, especially when so many different regimens have been used by the various labs. And second, all systems of the body are affected by food and feeding habits, so teasing out the criss-crossing lines of cause and effect is like driving through a hazardous landscape and trying to pick out the road in thick fog. Pankaj Kapahi and his colleagues, now at the Buck Institute in California, were among the first to home in on the nutrient-sensing networks – the cascade of events set in train by signals telling the cells what food is available and how the body should allocate its resources: whether towards protective mechanisms of recycling, repair and stress resistance, or to potentially damaging processes of growth.

At the centre of this network is our old friend TOR, the 'target of rapamycin' that we met in Chapter 4 when I was looking down the microscope at Lynne Cox's awesomely rejuvenated senescent cells. 'Every time you eat something,' says Kapahi, 'your body has to turn that food into protein mass or something, right? That's the job of this molecule, TOR. If there's a mutant in this pathway, animals are smaller. And what's amazing about this pathway is that it's conserved from plants to humans, so it is a very key growth sensor. As a postdoc, I was like, well, if this is such a key sensor, what if we reduced its activity, would we live longer? And that's what we found ... We were able to show that when you restrict the diet of a fly, the mechanism by which it's extending lifespan involves TOR. And it's turned out to be true in multiple species now.'

Another factor that's obviously important to the effect of calorie restriction, says Linda Partridge of UCL, is 'not having this great train of nutrients coming through and generating metabolic products and so on that the body doesn't need and that it has to cope with – to detoxify, get rid of. It's almost certainly just a slower throughput of stuff for the body to deal with.'

What that stuff consists of, she believes, is also important. Her work with flies at UCL, as well as evidence from mouse studies in other labs, has, over recent years, challenged the long-held assumption that the effect of dietary restriction on ageing is down to the calorie content of the diet alone, and that the nutrient mix is unimportant so long as it provides everything that's required for healthy development and sustenance. By tinkering with the nutrients in addition to cutting calories, researchers have found they can further manipulate the lifespans of their model animals. In Partridge's flies, for instance, cutting back on yeast (the protein in their diet) yields bigger dividends than cutting back on sugar, when measured calorie for calorie. Though the calorie/nutrient-mix issue remains unresolved and still somewhat controversial, most gerontologists today use the term 'dietary restriction' (or DR), which covers all permutations of the strategy, rather than CR.

But back to the issue left hanging at the beginning of the chapter – that of sex differences in the response to CR seen in McCay's original rat studies of 1935. This is important because it flags the pressing need to consider possible gender effects when analysing the results of studies – and, down the line a bit, in efforts to develop drugs and other anti-ageing therapies. Partridge has been investigating this phenomenon in her flies, where she noted early on that, contrary to McCay's finding with rats, DR had a much greater effect on the lifespan of females than it did on that of males. So what was happening? She set two of the postdocs in her lab, Jenny Regan and Mobina Khericha, to find out.

First they looked for differences between male and female flies, in the way individual tissues aged over the weeks of their lives, for clues to what might mediate the effects of DR. What stood out dramatically was the gut. Under the microscope the orderly honeycomb appearance of the lining of the gut in the females began to break down into a chaotic mosaic of cells as they aged, with holes appearing, tiny tumours, and scar tissue from healed wounds. What's more, the guts of female flies became leaky as they aged. Regan and Khericha put a dye in their food and could see it passing through the gut efficiently in young females, but beginning to get into other parts of the body as time passed. In the males there was almost no change across the weeks: the lining of their guts remained intact; it looked little different at the end of their lives from that of young females, and the system didn't leak. 'A lot [of the difference] is probably to do with the fact that female flies eat a lot more than males do,' explained Partridge at a symposium on healthy ageing held in Gothenburg, Sweden, in August 2016. 'They're basically machines for producing eggs; they have to eat a lot to make eggs, so the gut's doing a lot more in the way of dealing with nutrients.'

Looking at flies that had been on a DR regimen, however, Regan and Khericha found that the gut lining of old females was much less of a mess than in their free-eating counterparts, suggesting that the two scientists were on the right track – that slowing the deterioration of the gut was the key to the DR females' longer lives. But to be sure, they ran an intriguing experiment made possible by the curious physiology of their tiny models. Fruit flies don't have sex hormones, so the animal's sex is determined by the chromosomal make-up of each individual cell, which sends it down one or other path during differentiation. This meant that, with some nifty genetic footwork, the researchers could make flies with a mixture of sexes – for the purposes of this experiment, male flies with female guts. But that's not all. The fly's gut has several compartments, and the two scientists left the top compartment of the genetically modified creatures unchanged to act as a control. This compartment should continue to behave like a male bit of gut, they reasoned, thus reassuring them that any differences they saw between the two sections

were real differences in response to their environment, and not the result of general disruption to the gut caused by the scientists' manipulation of its anatomy.

Sure enough, as their modified males aged, the control section of the gut remained intact, while the feminised section deteriorated just like in the regular females, and it became similarly leaky. Intriguingly, the modified males continued to feed like males – that is, they ate a lot less than females – but they still developed the female gut pathology. '[This] argues that there's something about the female gut that's pre-adapted to the fact that it's going to get a lot of food, rather than [the pathology being] an immediate response to the food,' commented Partridge.

By nature male flies have shorter lifespans than females, and this was cut further in the modified males. The immune system is less robust in males than females, and Partridge reckons the modified flies may die prematurely because they can't cope as easily as females with the deterioration of their feminised gut; they succumb to 'a nice double whammy'. DR, however, gives these mixed-up males the same advantages as proper females: it slows the gut's deterioration and gives the flies extra days of life.

Lovely experiments, but what do they mean for us? 'I think, like a lot of experiments, this one probably raises more questions than it answers!' says Partridge. 'Can the gut account for sex differences in the response to other anti-ageing interventions? Is there a sex difference in gut ageing in mice and in humans? (The gut's been surprisingly ignored as an important tissue that might mediate ageing)? And is it important for the effects of anti-ageing interventions generally, even ones that aren't sexually dimorphic?* So I think there's a lot of work to do here – particularly in mice – starting to look more closely at the gut.'

---

* *i.e.* that don't show differences between the sexes.

Meanwhile, two research groups in the US have in recent years reported the results of the DR studies with rhesus monkeys started decades ago. (Ageing research in primates is especial challenging because the monkeys typically live 35–40 years). One study showed no effect on lifespan, while the other did. 'But what was very clear in both was that almost every aspect of function was improved during ageing, and the monkeys stayed largely disease-free,' said Partridge.

Meanwhile, our own model primate Dean Pomerleau, ever mindful of the science, has modified his diet and lifestyle in light of the monkey findings, easing up on the austerity. 'Much of the science that's come out in the last few years tends to poke holes in the idea that, at least in higher mammals, you're going to get the kind of dramatic life extension that you see in rodents,' he says. 'And even some of the rodent data suggest that avoiding obesity – so a very mild calorie restriction – is enough to get them the vast majority of the benefits seen with sort of weak-at-the-knees calorie restriction.

'But the real trigger for me were the primate studies. The take-away message there was: maybe you don't have to do this quite so intensively to get the benefits.' And those benefits – the dramatic improvements in health in the sunset years of life – are the holy grail, as far as most mainstream gerontologists are concerned. The search is now on to find drugs that might mimic the effects of dietary restriction without the personal pain and disruption to social life that are the corollary of years of obsessive calorie counting and food fixation.

CHAPTER NINE

The immune system – first responders

As we saw briefly in an earlier chapter, one of the commonest denominators of all age-related diseases – from cancer, diabetes, furred-up arteries and arthritis to blindness, lung diseases and dementia – is inflammation. This is not the inflammation familiar to us as hot red skin, swellings and pus, but chronic, low-grade, below-the-radar type inflammation caused in part by an immune system that never quite switches off as it should between challenges. So great a contribution does this chronic inflammation make to wearing away the smooth-running machinery of life that Italian immunologist Claudio Franceschi dubbed it 'inflammaging' in 2000 – a term that carries with it the connotation of a silent killer.

Many things contribute to inflammaging. Senescent cells send out signals to the immune system to come clear them away – signals that are persistent as these dysfunctional old cells accumulate with age. We live, most of the time, in pretty good harmony with billions of microbes that inhabit our guts and help us break down and absorb food, especially starches. But as we age, our gut walls begin to weaken and develop holes, much as we saw in flies in an earlier chapter, allowing microbes to leak out into the bloodstream. These truant bugs are also thought to keep the immune system constantly ticking over. So too the debris from the processes of living, such as free radicals and other waste products of metabolism, that tends to build up with the years as the clearing machinery becomes less efficient.

Most of these free radicals are generated by the mitochondria, the batteries of the cell where energy is produced. But scientists point the finger at another interesting role for the batteries in causing inflammaging. Mitochondria are thought to originate from a bacterium that, eons ago, was engulfed by an oxygen-breathing cell, which failed to digest it. The mitochondrion, trapped inside the cell, used the food in the host's cytoplasm to produce abundant energy, and eventually a mutually beneficial relationship developed between the bacterium and the host that fuelled them both and enabled them to flourish. Today the mitochondria are such thoroughly integrated elements in our cells that they are accepted as 'self' by the immune system. However, when they spill out from ruptured cells, or are themselves damaged by age and the rough and tumble of life, they release molecules reminiscent of their bacterial origins, and these alert the immune system to what looks like a foreign invasion.

Adipose tissue, or fat, is a source of inflammation, too, which is one reason why being overweight is a threat to health. But even those who don't obviously put on weight as they reach middle age will be accumulating fat, says Janet Lord, an immunologist and Director of the Institute of Inflammation and Ageing in Birmingham, UK. 'As you get older there's a tendency for stem cells to start to become fat cells,' she explains. 'So you can be quite a trim person but you could still have lots of fat tissue spread throughout the body. It seems to be that the receptors on the surface of some cells change, so they can't receive the signals that tell them to, say, "make a muscle cell" or "make a thymus cell", and the default position seems to be "make a fat cell".'

Many cells in our bodies can produce cytokines, tiny signalling molecules that carry messages back and forth as part of the ongoing chatter between them. Inflammatory cytokines communicate between the immune system and the rest of the body. One of the biggest sources of these little messengers is skeletal muscle tissue, simply because there is so much of it in our bodies, and here there's a balance to be struck, says Lord. Inactive muscle pumps out pro-inflammatory cytokines, while moving muscle makes anti-inflammatory cytokines to maintain equilibrium. 'That's why exercise is so good for you – and why it's bad to sit for any length of time. There are now independent studies that show that how much time you spend sedentary is an independent risk factor for ill health. So you can go out, as I do, for my morning run. But if I then sit for 10 hours, I might as well not have bothered. It wipes the benefits of my morning exercise out.'

On the strength of the evidence, Janet Lord has bought herself a standing desk because standing loads muscles and keeps them active, whereas they're not doing very much when you're sitting down. No one yet knows where the healthy balance lies between active and inactive muscles, and her team in Birmingham is running some studies to find out. They are working with elderly people in a care home where the residents typically sit for most of the day. Some of them are being asked to stand up for 10 minutes every hour, even if it means simply hanging on to a Zimmer frame. The inflammatory markers in this group are being compared with those of another group in the same care home who are carrying on with their normal sedentary routine.

The researchers are also working with a group of healthy, active elderly people, getting them to sit for varying periods of one, two or four hours, and looking to see at what point the pro- and anti-inflammatory signals get out of balance and inflammation sets in. 'All the medical guidelines at the moment just say "Try to be less sedentary", because we don't know any more than that,' says Lord. 'We need to be able to give people specific advice, and say, "Get up every hour" or "Don't sit for more than two hours". And again, we can give this advice to care-home managers to say, "Don't let your old people just sit there all day".'

Besides causing inflammaging by being permanently switched on at a low level, the immune system itself ages with time. Boiled down to basics, the job of our body's defence machinery is to detect and kill microbes such as bacteria and viruses; build specialised weapons (antibodies) against pathogens to which we are repeatedly exposed (this is immune memory); and to remove damaged or aberrant cells such as cancer cells from our bodies. The evidence is that the immune system gets weaker on all fronts as time goes by, and less efficient at doing its job.

This chapter will look at the 'first responders' to injury or infection – the paramedics of the immune system, if you like. Important among these are the dendritic cells that act as 'sentinels'. They sit just below the surface of the skin and mucous membranes – the areas of our bodies in closest contact with the outside world from which invaders will come – and their main purpose is to alert and activate the more specialist cells of the immune system when they sense danger.

I will discuss the dendritic cells at greater length in the next chapter. Here, I am going to concentrate on the dominant white blood cells among the first responders, the neutrophils, which home in on sites of injury or infection, and have a number of strategies for neutralising pathogens. Neutrophils can swallow the pathogens up (a process known as phagocytosis) and kill the engulfed bacteria with poisonous secretions. They can release molecules that kill bacteria in the immediate vicinity that haven't been swallowed, and signal for reinforcements from other cells in the immune system to come to the site of infection. And they can throw out sticky NETs (neutrophil extracellular traps) made from strands of DNA material designed to snare pathogens as they pass and stop the spread of infection. 'NETs in many ways are an equalizer, allowing the relatively slow moving cells of the immune system to "catch" highly motile or circulating bacteria, basically turning neutrophils into spider-like predators; setting traps and waiting for the prey to come to them,' write Craig Jenne and Paul Kubes of the University of Calgary, Canada.

In older people neutrophils are not as efficient at swallowing and killing – nor at casting their DNA nets – as they are in the young. But as we saw briefly in Chapter 4, one of the biggest problems with these immune cells is that they lose their sense of direction. When responding to a pro-inflammatory signal, elderly neutrophils zigzag through the tissue towards the site of injury like an emergency crew with a faulty GPS, causing collateral damage as they go. Looking at migratory patterns of neutrophils and their impact on tissues, Lord's team in Birmingham have found that 'even in healthy elders, the amount of damage caused by neutrophils just wandering around the body looking for infections is double that in a young person'. So when in life does this become a problem? 'We can pick it up in most 40- and 50-year-olds,' says Lord. 'But by the time you're 60 or 70 years old it's really bad. You struggle to find a neutrophil that is moving in the right direction efficiently.'

Blundering neutrophils are slow to reach their target, which is one reason why wounds heal so much more slowly as we age than they did when we were kids scuffing knees and elbows in the playground and quickly forming scabs. And they're one reason, too, why old people respond so poorly to infection. In severe infections such as pneumonia, old people's neutrophils are even more disorientated than usual, and the collateral damage they do in migrating to the site is vastly increased – up to five times higher than in a young person – and a potent cause of general frailty. Young people's neutrophils are also somewhat disorientated in cases of severe infection, but whereas they quickly return to pre-pneumonia levels of efficiency, the neutrophils of old people can't reset their GPS, leaving them vulnerable to repeat bouts of infection.

The problem with aged neutrophils, however, goes beyond faulty migration and difficulty killing off pathogens: in older people these cells are often extremely sluggish to respond to pro-inflammatory signals in the first place. Why? It turns out this is because of the chronic underlying inflammation which means the neutrophils are already activated and are unable to 'hear' the new signals clearly above the background noise. This discovery gave Lord and her colleagues an idea. Her lab had worked out the pathway between signal and response and knew there were drugs that could target this communication channel. The drugs in question were statins – already used by millions of people to lower cholesterol and protect them against heart disease, and which Lord and her colleagues had discovered also happen to improve the navigational ability of tired neutrophils. What if they used such a statin to interrupt the persistent activation of neutrophils in elderly people and restore somewhat their sense of direction – would it enable them to respond more effectively when a serious new infection set in?

They tested it on neutrophils in the lab and it worked. So they tried it in a small group of healthy older adults and found that just two weeks on statins made their neutrophils as good at navigation as a young person's. And they had the same effect on the neutrophils of elderly patients with pneumonia. But what about people who were already taking statins routinely for other purposes? Lord and her colleague Liz Sapey wanted to know if they had a better chance than people who were not on statins of surviving pneumonia, and it wasn't difficult to find out. Their lab is in the unusual position of being embedded within the busy Queen Elizabeth Hospital in Birmingham, built in 2010, which sees around one million patients a year. It's a paperless hospital and everything about patients is recorded on iPads at the ends of their beds. Within days the student Lord sent off to check had returned with the news that yes, elderly patients admitted with pneumonia who were already taking statins were less likely to die of their infection than those who were not.

'We've known for a long time that, if you look at how much the cholesterol is lowered in people on statins and then their health benefits, there wasn't a really good correlation,' said Lord. 'So some people, you put them on statins and their cholesterol doesn't go down very much, in others it goes down a lot, but they all seem to benefit from the drug. We now feel it's probably more to do with the effects on the immune system, because statins lower the inflammation, improve your neutrophils, and they make [more specialised] T cells function better too.' Large-scale clinical trials are now in the pipeline to determine whether or not statins should be recommended specifically for beefing up the immune system as well as for cardiovascular health.

Breaking a bone is a traumatic event at any age, but for an elderly person it can be a killer: roughly one in four old people who break a hip will die within a year. Lord's team in Birmingham set out to discover why. As you'd expect from all we've seen so far, ageing immune cells don't mount a very efficient response to the injury, but it transpires that this is clobbered further by the effects of stress hormones. The first thing that happens when any of us breaks a bone is that we get a burst of the hormone cortisol into the bloodstream. Its job is to prime the body to deal with an emergency (part of the 'fight or flight' response), and it suppresses non-essential functions, including immune-system activity. Soon after the cortisol burst, the stress-response machinery boosts the levels of a hormone called DHEAS, which is an immune enhancer and which should swiftly restore the immune balance. In young people it does so, but not in the elderly: production of DHEAS peaks around the age of 30, declining thereafter, so the balance is not restored in elderly patients with broken hips. This leaves them at high risk of catching an infection – typically pneumonia, a urinary tract infection or one of the nasties hanging around hospitals.

Interestingly, however, Lord's investigation revealed that the real culprit here was not simply the imbalance between cortisol and DHEAS, but also the effects of clinical depression – a condition that affects more than a third of their hip-fracture patient, and is probably also caused by the hormone imbalance, since this is known to affect mood. Checking everyone's neutrophil activity, they found that in all patients, these vital little fighters were still able to seek out and swallow bacteria, but those in depressed patients were unable to administer the *coup de grâce* to the engulfed pathogens. Surprisingly, in those who didn't become depressed, immune suppression was not such a big deal in hip-fracture patients though everyone – depressed and non-depressed alike – became more frail following their falls.

But what if the trauma a person suffers is psychological rather than physical? Lord and her colleagues monitored the neutrophil activity of elderly people who had lost a loved one, and found that 'bereavement is as powerful a stressor as a hip fracture, and it persists. We followed bereaved older adults for a year and their immune systems stayed suppressed.' Recalling the stories one hears so often about a couple, married for 40 years or more, where one of them dies and the other soon follows, she commented, 'They go down with infections; it's nearly always pneumonia. So I always say they die not of a broken heart, but of a broken immune system.'

CHAPTER TEN

The immune system – the specialists take over

So far I've been talking about the *innate immune system* – the first responders with their broad-spectrum activity. Here I'm going to look at the other, more 'intelligent' branch of the body's defence machinery, the *adaptive immune system*, responsible for building resistance to foreign invaders that might threaten us repeatedly, and at what happens to it as we age. This arm of the immune system only exists in vertebrates (that is, creatures like us with a backbone and braincase), says Janko Nikolich-Zugich, Professor of Immunobiology at the University of Arizona, US. 'It has evolved specifically to defend multicellular organisms that are very complex, and it's very highly microbe-specific. It is laser-like in its action, and it orchestrates elimination of the microbe in a very, very precise manner.' It does this by developing specialist weapons designed to recognise and neutralise every individual foreign body that we encounter.

The cells of the adaptive immune system are the B cells, made by the bone marrow, and the T cells, made by the thymus (hence B and T). The B cells make antibodies – individually tailored weapons – in response to invasion by specific bugs. These circulate in the bloodstream and lymphatic system ready to fight the bug if and when it returns. The T cells, on the other hand, when called upon to fight an invader (typically a virus that gets right inside cells), turn some of their naïve cells – that is, ones that are not yet specialists – into heavily armed killers that can proliferate enormously to produce hundreds of thousands of copies of themselves. These killer cells swarm to the attack, slaying the cells that are infected by the virus. Once the invader is vanquished, the majority of these specialist soldiers die, but a few survive and return to barracks with the memory of the invader still intact. These 'memory T cells', as they're called, are ready to spring into action again the moment they recognise a repeat invasion by a bug, churning out millions of copies of tailored killer cells once more.

This is the basis of immunity and vaccination, and so efficient is the adaptive immune system that we're not even aware of reinvasions. That is, until we get old – for as we have seen, every part of our body's defence machinery ages too, and small defects that develop all along the chain of command compound one another to produce increasingly serious immune deficiency overall.

In the last chapter, I mentioned briefly the dendritic cells which act as sentinels and are responsible for activating the adaptive immune system when an invasion occurs. The dendritic cells do this by collecting bits of the invading organism, which they present to the B and T cells, the soldiers of the adaptive immune system, waiting in barracks in the lymph nodes to be called up. The B and T cells then design their specialist weapons for attack accordingly.

In his lab in Arizona, Nikolich-Zugich studies the interaction between the sentinels and the adaptive immune cells. As they age, he says, 'the dendritic cells, when we examine them carefully, are showing reduced uptake of the bug. That makes the T-cell priming very wimpy, because [the dendritic cells] haven't been really activated to a sufficient extent to present enough molecules of the bug to the T cell. And they're also not making all of the other juicy factors and molecules that the T cell needs to see in order to be kicked into activation.'

So the sentinels in elderly people are doing a pretty pathetic job of gathering intelligence. In turn, the populations of adaptive immune cells (both B and T) become depleted from repeated stimulation and proliferation over the years, and the cells' telomeres shorten, ultimately leading to senescence. There are not, therefore, as many naïve, uneducated cells waiting in barracks to respond to new threats. Inadequately stimulated and ageing B cells make poor-quality antibodies, while ageing T cells struggle to produce educated new soldiers, and the small reserves of specialist T cells in the army begin to lose their memory, leaving us ever more poorly defended against invaders. Sometimes old enemies take the opportunity to renew their attack, as I myself discovered.

I was nine years old, and my parents had rented a caravan in Ireland for a family holiday. It was a stressful time: my father, a doctor of tropical medicine, was between assignments. We had recently returned from three years in Borneo, and were unsure where dad's next posting would take us. He developed an angry boil on his neck, and two days into the holiday my younger sister came out in a rash – big weepy spots all over her body. My older sister and I followed suit a few days later. We had chickenpox. I seem to remember the caravan was a tight squeeze for four sick people and one overburdened one, and it rained for two weeks: not the happiest of family holidays.

Fast-forward four decades and more. I am sitting on a plane headed for Belgrade where I am to report on HIV/AIDS for the World Health Organization. I am restless and uncomfortable, with a strange, nervy pain around my middle, an outbreak of tiny blisters and skin that is sensitive to the touch. I have shingles. What links these two distant bouts of ill health is the varicella zoster virus that causes chickenpox, and that can remain dormant in nerve cells after the initial infection, kept in check by a healthy young immune system. As this weakens with age or stress, immune surveillance wanes, and the virus can break free of restraint and travel to the nerve endings in the skin to produce shingles.

A more predictable threat to us as we get older, however, is the winter flu season. And this virus delivers a double whammy: not only are elderly people especially vulnerable to catching flu, but their response to vaccination – which works by priming the adaptive immune system to produce its specialist weapons – is often poor because of the ageing defence machinery. Furthermore, the protective effect of vaccination doesn't last as long as in younger people because of the fading memory of the cells. People over 75 are almost twice as likely as those below 65 to catch flu when it's around, and one in three of those that do will end up in hospital with a life-threatening illness. 'In the US alone, about thirty to forty thousand people die every year of flu,' says Nikolich-Zugich. 'That's only one infection that affects older adults, and over 95 per cent of the people that die are over 65 years of age.'

Few people below the age of 65, however, will end up in hospital or die from flu. That is, unless they are practising caloric restriction, according to Linda Partridge of UCL, whom we met in Chapter 7. One of the downsides of this dieting strategy, which has proved tantalisingly effective at slowing down the ageing process in so many species, including us, is that it suppresses the immune system. And out in the real world, beyond the controlled and bug-free environment of the lab, this can be fatal. 'DR animals are usually less good at clearing viruses,' says Partridge. 'The one that's been particularly looked at is flu. And that's the same in humans who voluntarily dietarily restrict themselves – if they get flu they're in trouble.' People on DR are also less good at healing wounds, she says. 'If there's any question of trauma or infection they have to refeed.'

The good news is that there are things we can do to delay the decline in our body's defences. Lifestyle plays a crucial role. Lord and her team in Birmingham recruited a group of people aged 55–80 who had been cyclists most of their adult lives – 'Not just "I'm going to get on my bike to go to the shops" type cyclists but the kind you see in their Lycra on a Sunday,' she says. Over three days in the lab they subjected the recruits to a battery of fitness tests, measuring everything from heart, lung, cognitive and immune function to muscle mass, bone density and blood flow to the brain. While muscle and bone showed little sign of ageing, heart and lung function did decline somewhat with age. But the effect of this tough regular exercise on the adaptive immune system was remarkable.

The T cells arise, like all other immune cells, from the pool of stem cells in the bone marrow, and they're sent to the thymus to be educated into specialist defenders. The thymus – a gland about the size of a plum that sits just above the heart – is at its largest during childhood, when it's busiest creating an army of cells to attack foreign invaders. After puberty the thymus begins to shrink and is progressively replaced by fatty tissue. Production of new, specialist T cells declines progressively too. But in Lord's cyclists this wasn't the case. 'They had really good thymic output of new T cells,' she said. 'The levels we measured were as good as a 20-year-old's.' What she thinks is happening is that the cyclists' muscles are much more efficient than those of their less active peers at maintaining levels of the growth hormones that keep the thymus going.

To be included in her study, the cyclists needed to be able to do 100km (62 miles) in 6.5 hours for a man or 60km (37 miles) in 5.5 hours for a woman, and to have met these targets at least three times in the previous fortnight. But does one have to keep to such a strenuous routine to reap the benefits? The Lord team looked at the immune function of 200 others, some of whom were meeting the current NHS guidelines for physical activity and some of whom were regular couch potatoes. The active group had much lower levels of chronic inflammation than the sedentary lot, though all had shrunken thymuses, reflective of the normal decline with age. 'You probably do have to do quite a good level of exercise to protect the thymus,' says Lord. But as far as maintaining the balance between pro- and anti-inflammatory signals from the muscles during the normal routines of daily life is concerned, 'You get the biggest bang for your buck by simply changing from being sedentary to doing a little bit of exercise.'

CHAPTER ELEVEN

The bugs fight back

Clearly the inexorable decline of the immune system with the passing years is a key driver of age-related disease. But what about the other way round? Do non-age-related diseases that hammer the immune system also have an effect on the rate of ageing? Let's look at cytomegalovirus. CMV, as it's commonly known, is a herpes virus in the same family as the one that causes chickenpox and shingles. It's enormously widespread in the world, with up to 90 per cent of us likely to be infected by the time we reach old age. It's carried in bodily fluids: saliva, blood, urine, semen, vaginal fluid and breast milk.

'If you get CMV in utero,' says Nikolich-Zugich, 'it's actually quite devastating' – a potential cause of deafness, blindness, mental impairment and cerebral palsy. Most of us, however, pick it up as little kids in the rough and tumble of the school playground, or else a bit later when we become sexually active. And, ten to one, we will be totally unaware of its presence because, typically, it causes no outward sign of disease in healthy people. But this is a huge virus that, once infected, we carry for life, and it has a profound impact on our immune systems as we grow old, potentially increasing the risk of cardiovascular disease, Alzheimer's, some cancers and general frailty.

On initial infection with CMV the immune system launches a mighty attack to subdue the virus, which holes up in the stem cells of the blood – the progenitors of all the specialised cells of the immune system – as well as infecting several of the specialised cells themselves. This means CMV lives effectively in the very heart of our defence machinery for the rest of our lives. But it costs the virus dearly to do so.

'CMV devotes about 90 per cent of its DNA, or more, to negotiating this very delicate balance with the host,' says Nikolich-Zugich. But we in turn devote a huge amount of our immune resources to keeping the virus in check by creating armies of killer T cells whenever it threatens to break out again, and each time adding to the pool of memory T cells able to recognise the virus instantly should it do so. Over time our response to CMV comes to dominate the system, with around 10 per cent of all T cells devoted to suppressing this one virus, and up to half of the supply of memory cells being CMV-specific in some elderly people – a phenomenon known as 'memory inflation', which may leave fewer resources for fighting other infections.

Besides targeting the blood cells when it invades, CMV also goes for the cells lining the walls of large blood vessels, where the fatty plaques of atherosclerosis – a typical age-related disease – build up, and where CMV is often to be found. But here it gets really hard to distinguish between cause and effect. Is the virus helping to create the life-threatening plaques on the artery walls? Or is it simply carried there by the immune cells answering the emergency call to the site of inflammation – the very cells which, as we've just seen, are the prime hiding places of the latent virus? In other words, is the virus causing direct trouble in the arteries or is it just an innocent bystanders? This remains one of the big unanswered questions about CMV as a possible driving force in ageing.

And there are a number of others, too. What, for example, prompts the virus to reactivate? Is it triggered by something in the host, or something intrinsic to the virus? Is CMV ever truly latent, and causing inflammation only when it reactivates? Or is it smouldering, leaking out viral particles all the time that stimulate constant, low-grade inflammation and contribute directly to inflammaging?

None of these questions is easy to answer, partly because there are no really good animal models for CMV infection in humans. The virus is species-specific, tailoring itself minutely to its host, so that you can't infect a mouse, or even a monkey or chimp, with a human strain, or vice versa. On the plus side, CMV's behaviour in the different species is thought to be very similar. But on the minus side, some of the everyday things that are thought to stimulate reactivation of the virus in us humans, such as stress, are hard to reproduce faithfully in model organisms in order to test the hypotheses.

'We've been working on CMV and ageing in mouse models for a while,' says Nikolich-Zugich. 'But we've concluded that one of the big problems with this is that mice, in the way we're treating them, lead pretty stressful lives anyway. One of the worst stresses [for the mouse] is when a graduate student suddenly pops up in the room and tries to grab him by the tail! In humans we know – and we've been able to measure it, actually – that every time there is another little infection, or you're upset at something, CMV reactivates at some level.' They now have to mimic this in mice by keeping them in much more human-like conditions where they can manipulate the levels of stress up and down.

Even so, it's hard to know how much of what you discover in a mouse is ever applicable to a human being. But if you read between the lines, a recent study with twins by a group at Stanford University, California, offers intriguing evidence that CMV, rather than simply being a ubiquitous and meddlesome diversion, may be actively contributing to the process of ageing through its influence on the immune system.

The study – part of the endlessly fascinating nature/nurture debate – was not set up to look at ageing per se but to assess the contribution our genes make to how well our immune systems work. Among 105 pairs of twins aged between 8 and 82 years, the team recruited 78 pairs who were identical – ideal subjects because any differences between them in whatever was being measured could be readily attributed to environmental factors, since the twins, having developed from a single egg, shared their genetic make-up. The scientists looked at a total of 204 variables – 204 individual elements in the mighty machine that is the immune system – including populations of some immune cells, and signalling and other important proteins circulating in the blood. They found that the primary influence on the functioning of virtually everything they measured was environmental rather than genetic. What's more, the older the pair of twins was, the stronger the effect of the environment and the weaker that of their genes on the behaviour of their immune systems. The researchers concluded that the most likely environmental influence would be the many different microbes – bacteria, viruses, fungi and the rest – a person had been exposed to during their lifetime. Naturally, these would accumulate with age, helping to explain why the twins' genetic heritage was less and less important as they went up the age range.

One microbe offered a particularly striking example. 'Among some of the twin pairings,' says Nikolich-Zugich, 'they had couples in which one had CMV and the other did not, and that was absolutely stunning. It showed that CMV positivity modulated more than 50 per cent of all of the immune parameters they measured. It was an enormous, enormous modulator of everything that happened.'

But as if to confound all that's been discovered so far – and to remind researchers that there are few certainties in science – evidence is emerging that CMV may sometimes be beneficial for its carriers. 'It may even improve immune defence against other infections, as well as the response to vaccination, in an adult population,' says Nikolich-Zugich. 'So understanding the intricacies of the dance between this virus and our bodies is really important to improving immune ageing.'

Most of us, of course, are unaware of our CMV status, and whether or not our immune systems are engaged in a ceaseless, silent battle with a virus that may be hastening our decrepitude. But that's not the case with HIV: this microbe makes its presence known only too clearly in its swift destruction of our body's defences, or else in the lifelong reliance on powerful drugs to keep it at bay. So what is the AIDS virus doing to infected people as they age?

CHAPTER TWELVE

HIV/AIDS – adding insult to injury

I have a deep personal interest in AIDS. As a freelance writer I happened to be working on assignment at the World Health Organization (WHO) in Geneva when the first mention was made of a strange new phenomenon that was baffling medics – a cluster of young, gay men in San Francisco coming down with a rare type of pneumonia typically seen only in elderly people or others with drastically weakened immune systems. The report appeared in a dry little publication, the *Morbidity and Mortality Weekly Report* or *MMWR* – just a collection of world health statistics really – put out by the Centers for Disease Control in Atlanta, US. This was 1981, and I was subsequently asked to write a short piece on the mystery disease for one of WHO's journals. No one had any inkling then of what we were facing – a devastating epidemic that would sweep the world; the spread of a virus that was, in the early days, a death sentence for all who contracted it.

I spent the next 20 years reporting on the AIDS pandemic from the front line, mostly in Africa, but in other countries and continents too, as the virus spread like a bush fire. I remember the villages on the shores of Lake Victoria in Tanzania and Uganda where the busiest workers were the coffin makers, knee-deep in sawdust in their small roadside workshops. I remember too the fresh graves dotted like raw wounds among the banana palms in everyone's back yards. Sometimes five, six or more members of a family were wiped out – an entire generation of young adults – leaving children to be raised by grandparents deprived of care themselves in their old age.

And I remember the heroic effort by scientists to understand the virus and develop weapons against it; and the triumph eventually of antiretroviral therapy that has made HIV infection today a condition to be lived with rather than to die from for those who are fortunate enough to have access to the drugs. Today, more than 60 per cent of people with HIV in San Francisco, and over half of those in New York City, are over the age of 50 – survivors of the pandemic that ripped the fabric of their communities, and killed their lovers and friends in huge numbers. The comparable figure for the UK is around 34 per cent of people accessing treatment for HIV, and here, in 2015 alone, more than 1,000 people over 50 years old were newly diagnosed with the virus. But the sting in the tail of the survival stories is that HIV seems to accelerate the ageing process. Today people living with HIV – even when the virus is tightly controlled by drug therapy – are developing typical age-related diseases 15 to 20 years earlier than their non-infected peers.

Peter Hunt, an AIDS researcher at the University of California, San Francisco, became interested in HIV as a medical student at Yale in the mid-1990s. By then the virus first reported in California had spread around the globe; it was decimating communities in many parts of sub-Saharan Africa and South-East Asia, and had already killed more Americans than the Korean and Vietnam wars combined. Here was a disease with social, psychological and political dimensions as well as intriguing biology – an exciting challenge for an aspiring young medic. Hunt was hooked. Besides, by the time he was in med school, the first generation of really powerful antiretroviral drugs, or ARVs, was appearing in the clinic, transforming the prospect for saving the lives of people who caught the virus and who had, so recently, been without hope of survival.

Moving to California after qualifying as a doctor, Hunt focused on treating patients with HIV/AIDS, but found himself more and more drawn to research. Why, he wondered, did the immune systems of his patients fail to recover properly even when the virus was successfully controlled by the cocktail of drugs they were taking? Today he spends most of his time in the lab, but he continues to see patients and says the drug treatment has become so sophisticated that 'in many people, HIV infection is a chronic condition like high blood pressure'. In recent years, however, he and his colleagues have noticed a new phenomenon: people who are being successfully treated with ARVs showing signs of premature ageing. They are coming to his clinic with a range of conditions from cardiovascular disease, diabetes and osteoporosis to lung, liver and kidney diseases and declining mental powers, much earlier than would be expected.

Others had noticed this trend too, and for some time the assumption was that the premature risk of these age-associated conditions in HIV-positive people was down to the toxicity of the powerful ARVs. Much effort was spent on trying to develop safer and more precisely targeted drugs. Then in 2002 an international consortium of scientists set up the SMART trial (standing for Strategies for Management of Antiretroviral Therapy), designed to see whether interrupting treatment periodically as the immune system rallied would reduce the side effects. Doctors were chiefly concerned about cardiovascular problems and metabolic complications, including abnormal distribution of body fat and heightened risk of diabetes in their clients on ARVs. 'The idea,' said Hunt, 'was only to use the drugs when you absolutely need them. They have all these toxicities, so maybe we can have this drug-sparing strategy to minimise the toxicity while still giving the immune system enough of a boost.'

Treatment decisions are based on the state of a person's immune system as measured by the number of CD4 T cells per cubic millimetre of blood – your so-called 'CD4 count'. Anything between 460 and 1,600 is considered normal, and at the time of the SMART trial the customary practice among doctors caring for HIV-infected people in the US was to recommend to anyone with a CD4 count below 250 that they start treatment.

The trial recruited 5,472 HIV-positive people from 318 sites in 33 countries. Some were already on ARVs, others were not, but to be eligible for inclusion in the study a person had to have a CD4 count above 350 at the start. Participants were then randomly assigned to one of two treatment strategies: one group either continued with, or started, antiretroviral therapy, which they were to stay on continuously, as is the common practice. The other group was to interrupt treatment if they were already on ARVs, or defer treatment until their CD4 count had fallen below 250. At this point they were to resume treatment or start for the first time, and carry on until their CD4 count had recovered to more than 350, when they were to take a 'treatment holiday' until their CD4 count had dropped again to 250 – the red light for going back on drugs. And so they continued, either on uninterrupted treatment, or taking drug holidays as their CD4 cells rallied and taking up with the drugs again as their immune systems declined. Over the course of the trial those on the intermittent regimen, as a group, consumed about one-third of the quantity of ARVs consumed by those on continuous treatment.

All the evidence to date was that a count of 250 was a safe threshold – a level at which an infected person's immune system was still strong enough to keep AIDS at bay. Furthermore, research findings suggested that disease and death in HIV-infected people with CD4 counts higher than 250 were usually caused by the drugs themselves, or something else unconnected to their viral infection. Thus the scientists running the trial around the world were pretty confident that not only would those having regular respite from the toxic drugs enjoy a better quality of life, but they would experience fewer of the diseases they attributed to the drugs than those on regular treatment.

'But then,' says Hunt, 'the trial proved the exact opposite: that it was the people who were interrupting therapy that seemed to have an increased risk of heart disease, cancer, liver disease, kidney disease. So all of a sudden the whole field woke up to say, "Gosh, you know, the virus is worse for you than the drugs."' From his vantage point on the front line of treatment in Californian clinics, Hunt, for one, was not surprised. 'I was sceptical that the drugs explained everything, because as we learnt about toxicity, new drugs were developed, and we would always put our patients on those that had fewer toxicities. I was doing that myself, but it just seemed that that wasn't enough ... There seemed to be something else,' he says. 'Since SMART there has been an explosion of research examining each one of these age-associated morbidities to see whether it's increased in HIV infection, and many of them are. Not all of them, but many of them are. And research is also trying to understand the biologic mechanisms that might explain that.'

Once again an accusatory finger points at inflammation, and here there seem to be two key drivers of the process in people living with HIV. 'There is HIV itself, which continues to leach out of infected cells,' says Hunt. 'All of our drugs block new rounds of replication, but they don't block release of virus from infected cells – that continues.' Even when someone's viral load is undetectable in the blood, HIV is still able to shed particles in the lymph tissue, which is where it holes up indefinitely, largely beyond the reach of ARVs.

Then there's the problem of leaky guts. As we've already seen, this is believed to be a cause of inflammation in 'normal' ageing too, when bits and pieces of microbes are able to escape into the bloodstream from their rightful place in the intestines. But in people with HIV the problem is accentuated – and the association with ageing clear – because breakdown of the lining of the gut is one of the earliest events following infection with the virus, no matter what age that occurs. 'We're learning a lot about the mechanisms by which HIV establishes an initial infection and long-standing reservoir in people,' says Hunt. 'One key feature is that regardless of how you get infected – whether it's through sex or intravenous exposure – the virus finds its way to the gut, and that's how everything starts.' HIV has evolved mechanisms for homing in on the cells of the gut, which are particularly hospitable to it, and where there's an initial explosion of virus production to flood the body. 'This is something very central to the pathogenesis of HIV,' explains Hunt. 'And then it leaves behind all this damage that may not be fully repaired with treatment.'

Besides looking at the causes and role of inflammation in the premature ageing of people with HIV, Hunt is investigating how the damage done by HIV to the immune system itself might be playing a part. His main focus is the CD8, or killer T cells. Under normal circumstances, these cells, when called upon to fight a virus, go into a frenzy of division and differentiation to pump out an army of specialist killers. As we age naturally, the pool of naïve CD8 cells – that is, the ones waiting to be trained up – progressively declines, to be outnumbered by the already-trained specialist killers, which are not so good at responding to new challenges. In people with HIV, however, the killer T cells seem to get stuck midway through their differentiation from naïve cells to specialist killers, and they fail to proliferate.

The question preoccupying Hunt and his colleagues is, what is the function of these cells which have failed to develop fully? This is still a mystery, though the presence of lots of these arrested cells in people with HIV seems to set the scene for the diseases of old age and is a strong predictor of death. 'We think this reflects an important difference between what's happening in HIV infection and what's happening in ageing,' comments Hunt. 'Both scenarios probably reflect a defect in immune function in the body, but the way you get there is very different.'

So if the virus itself is the most likely driver of premature ageing in people living with HIV, rather than the powerful drugs which were the prime suspects for so long, when is the best time for an infected person to start treatment? This was the obvious question arising from the revelations of the SMART trial. At the time, the recommendation from the World Health Organization was that treatment should be initiated as soon as possible once a person's CD4 count had fallen below 200 – the point at which the immune system really begins to struggle to fight off the infections and diseases that characterise AIDS, though these can still break through the immune defences at higher counts. In reality, even this was an ambitious target: when I was reporting on the effort to provide antiretrovirals in some of the worst-hit countries of Africa and Asia in the early 2000s, people were very often near death, with CD4 counts below 10 and already suffering from AIDS, before they could get hold of the drugs. The effect of treatment was often near-miraculous, allowing people who had been bedridden and hopeless to return to their normal lives.

But today it's these people, who started on ARVs at the lowest threshold of immunity, who appear to be at greatest risk of the panoply of age-associated diseases, even if their virus is under control and their CD4 count once again strong. They're the ones turning up at clinics across the world with cardiovascular problems, lung, kidney and liver diseases, diabetes and cancer years before they would normally expect to be at risk. Over the years, the recommendations and general practice in HIV clinics have changed in the light of experience. The threshold for treatment has risen to a CD4 count of 350 in many places, but the guidelines have been inconsistent and based largely on the shaky foundations of observation. So in 2009 an international network of researchers set up the Strategic Timing of Antiretroviral Therapy, or START, trial to measure systematically the relative pros and cons of beginning treatment as soon as a person is diagnosed with HIV, at a CD4 count of 500 and above, compared with deferring treatment in a newly diagnosed person with a similarly high CD4 count until this has dropped to 350. For the individual person who becomes infected, the implications are huge: ARVs are a lifelong commitment and the drugs do still have a range of unwelcome, more or less serious side effects besides those that are life-threatening, such as the abnormal distribution of body fat.

In all, 4,685 people aged 29 to 44 years, who had been diagnosed with HIV but had not yet begun treatment, joined the START trial, which involved 215 clinics in 35 countries. Participants, all of whom had CD4 counts above 500 at recruitment, were divided roughly equally between the two treatment strategies. Six years later, nearly half of the 'deferred treatment' group and virtually all the 'immediate treatment' group were on ARVs. At the point of starting treatment, the viral load – the number of virus particles in a millilitre of blood – of those in the deferred treatment group was, on average, more than three times higher than that in the immediate treatment group. Unsurprising, perhaps, since their bodies' defences were a lot weaker by then than the defences of the immediate starters. After a year of the drugs, however, the virus had been fully suppressed, meaning it was virtually undetectable in the blood, in almost everyone, regardless of the level of circulating virus at the start.

The study followed participants for an average of three years, and outcomes other than viral suppression soon made it clear that there were big advantages to starting therapy without delay. By early 2015, those who had deferred treatment were more than twice as likely as the immediate starters to have suffered a serious illness, either AIDS-related (most commonly tuberculosis, Kaposi's sarcoma or non-Hodgkin's lymphoma) or non-AIDS-related (commonly another type of cancer, a heart attack or death from some other disease). Looking at AIDS-related conditions alone, those who started treatment immediately had a 70 per cent lower chance of getting sick than the deferrers. So clear were the results that the study was stopped some 18 months early to allow those who were not yet on therapy to begin taking the drugs with no further delay.

The lesson from this study and his own experience on the front line is that 'the disease state at which you start therapy makes a huge difference to HIV,' says Hunt. 'There seems to be a point of no return. Your CD4 count may get back to normal levels if you start ARV therapy late, but you still may be at much higher risk of multiple different diseases moving forward. Even at very early stages of infection, delaying [treatment] just a little bit seems to have an impact.'

In the light of the START trial results the World Health Organization has revised its guidelines to recommend that the threshold for starting treatment be raised to a CD4 count of 500. But the benefits of any new treatment practices will be a long time coming. At the end of 2016, just over half the people living with HIV worldwide were on antiretrovirals, huge numbers of others were believed to be living with the virus but unaware of their status, and the great majority of those on treatment will have started taking the drugs at late stages of infection, when their immune systems were already seriously damaged.

Not everyone is convinced by the case for premature ageing among people living with HIV. Some in the HIV/AIDS community, both scientists and activists, believe it's been exaggerated – or even misinterpreted. It's near impossible, they contend, to separate the effects of the virus from the effects of lifestyle on biology. But, says Hunt, 'there are a number of very well-done studies recently that have tried really carefully to match people with HIV to people without HIV for other behavioural factors – you know, smoking, number of sexual partners, whether they use drugs, whether they drink, all of those things. In some of the studies, where you carefully control for those factors, the amount of inflammation you see in HIV-infected people is not quite as great as in studies that don't match controls so carefully. But even in those studies, if you look at disease outcomes – you know, are you getting heart disease, cancer? – there still appears to be an increased risk. And if you look at the [characteristics of ageing more generally], using multiple biomarkers, there still seems to be a big difference – a *significant* difference – between people with HIV and really well-matched controls.'

As with so many health issues, he points out, we're looking here at a world divided between rich and poor, between 'the haves' and 'the have-nots'. 'A lot's been said lately that the life expectancy really is getting close to normal for people living with HIV, and that is true among people who start treatment very early in the course of their disease. But it's not true for the vast majority of people around the world who have started treatment at late disease stages and are now growing older with HIV. For them life expectancy frequently is shorter – and probably by about two decades.' Nor should it be forgotten that a person's 'health-span' – their quality of life – will be eroded by the experience of living with the deadly virus, no matter how long they live.

CHAPTER THIRTEEN

Epigenetics and chronology – the two faces of time

What exactly do we mean by 'premature ageing'? Except in extreme circumstances, this is a hard concept to grasp, for once people reach maturity it becomes increasingly difficult to judge accurately, just by looking at them, how old they are as they become weathered by their own unique experiences of living. So the question is, does our biology continue to keep step systematically with our chronology once the developmental programme has run its course? If so, is it played out at the level of the individual cells, or the tissues, or our bodies as a whole? And how predictable is it?

For some fascinating insights, let's take a look at epigenetics – one of the newest frontiers in ageing research. Epigenetics means 'beyond the genes', and it refers to the chemical switches attached to our DNA that orchestrate the activity of our genes, turning them on and off as appropriate, and modifying their tone. In common with all living organisms, we humans have a basic epigenome – an 'instruction manual' that controls the function of the genes so that a huge variety of different cell types can be fashioned from identical dollops of DNA, acquired when sperm and egg first met at the dawn of our lives, and present in every cell.

But that's not all. The epigenome is also the 'missing link' between our genes and our environment, since the basic manual can be edited throughout life by the addition or removal of chemical switches in response to a wide variety of cues. This mechanism enables organisms, including us, to adapt swiftly and perhaps transiently to environmental conditions without any changes to our basic DNA – in other words, without waiting for the ineffably slow process of natural selection to fit us more fundamentally to our world.

Take a look at these striking examples from Nature. There's a species of locust that develops without functional wings if there is likely to be lots of food and little competition in the world into which it's hatching. But if the signals it receives in the egg suggest high population density and lots of competition, it will develop wings that can carry it far afield to forage. The two bodily designs have identical genes, but they look so different that for a long time biologists believed they were separate species. The same mechanism is at work in honeybees, which decide when they are still larvae whether to develop into a queen or a worker bee, based on population dynamics and predictions about the role they'll play in the colony. And there's a species of meadow vole that is born with a thicker coat in winter than summer, indicating subtle, environmentally sensitive differences in how the genes are played, as orchestrated by the epigenome.

Here, in essence, is how the mechanism works. Structurally, DNA is an ultra-fine ribbon of continuous genetic material arranged in the famous corkscrew shape, the 'double helix', discovered by James Watson and Francis Crick in 1953 based on the images provided by Rosalind Franklin. Each of the trillions of cells in our bodies (except the red blood cells, which are unique) contains about 1.8m (6 ft) of the stuff, and to illustrate just how ineffably fine this ribbon of data is, some geek has worked out that if a single person's DNA was stretched out end to end it would reach to the moon and back more than 3,000 times.

For packaging purposes, the DNA ribbon is wrapped around a series of core proteins, known as histones, like sewing thread on spools. These histone bundles are known as nucleosomes, and they are strung together like beads on a necklace and folded into a structure known as chromatin, which is tightly compressed to fit into the nucleus of the cell. The epigenetic tags, or 'switches', are attached either to the DNA directly in the links between the nucleosome beads, or to the histones (the 'spools'), and they have the effect of relaxing the compacted DNA so that the genes can be read and activated, or tightening its compression so that the genes are unavailable to the copying machinery and kept silenced.

The most widely studied of the epigenetic mechanisms is DNA methylation, which attaches chemical tags known as methyl groups to the DNA, thereby suppressing the activity of specific genes. The methyl tags can be removed by specialist enzymes, too, and during development of an organism from seed to maturity, DNA methylation is a busily dynamic process of adding and removing tags to orchestrate the genes for the normal growth and differentiation of cells. When we reach adulthood, our methylome (that is, the established pattern of DNA methylation in our cells) is much more stable.

However, researchers are discovering more and more subtle ways in which our epigenome as a whole can be affected by our environment – by such things as our dietary habits, exercise, exposure to pollution, smoking, and alcohol and drug use. It has long been known that ageing, too, affects the epigenome with, among other changes, the gradual loss of methyl tags and the laying down of new ones, some of which can be clearly associated with disease. If, for example, a new tag appears in a position on the DNA close to a vital tumour-suppressor gene, thereby silencing it, the risk of cancer is increased.

But despite the apparent plasticity and sensitivity of the epigenome to myriad different cues, both internal and external to our bodies, there is strong and intriguing evidence of some inexorable process going on behind the scenes that can be read in the pattern of our epigenomes. In 2013 Steve Horvath, a mathematician in the genetics department of UCLA, came up with a model for an 'epigenetic clock' that is able to correlate the biological age of a broad spectrum of cells and tissues throughout our bodies with our chronological age more closely than any other biomarker discovered so far.

This was a long and monumentally painstaking exercise in data crunching and mathematical modelling that involved analysis of the methylation patterns in 8,000 human samples of known chronological age, drawn from 82 publicly available DNA data sets. There are millions of potential methylation sites in our DNA, but Horvath finally identified 353 sites with a sufficiently consistent pattern of change across the years to reflect age, and shared by 51 different healthy tissue and cell types. He looked too at the methylation patterns in 20 different kinds of cancer to get a sense of how disease might affect the rate of ageing – the ticking of the epigenetic clock – in our cells and tissues.

Horvath's results were remarkable: overall, his biological clock was able to estimate a person's chronological age accurately to within 3.6 years on average. But the correlation was even closer in some specific cell types. Saliva, for example, gave a prediction to within 2.7 years, some white blood cells to within 1.9 years, and brain cells to within 1.5 years. As anticipated, the epigenetic clock in embryonic stem cells registered close to zero. By contrast, the clock showed a huge imbalance between biology and chronology in the cancerous tissue samples, which were an average of 36 years older than the people from whom they had been taken. Across the 20 different tumour types Horvath tested, however, there were wide differences – with the least pronounced disparity (or the closest correlation) between biology and chronology being in the cancers associated with mutations in key genes such as the tumour-suppressor gene p53. This intriguing observation added strength to Horvath's theory about what might be driving the epigenetic clock.

In his 2013 paper in the journal *Genome Biology*, he describes how the clock ticks fast during the period of dynamic growth from embryo to adult, and then slows down and plateaus out after maturity. He suggests this reflects the energy expended on maintenance of the epigenome to ensure its stability at a time when the faithful orchestration of the genes responsible for development is critical, and cellular systems are under greatest stress. Once we reach maturity, the pressure on performance is eased and our bodies cut back on investment in epigenetic maintenance. (Are you hearing echoes of Tom Kirkwood's disposable soma theory?).

So how do the variable ticking speeds of the clock in cancer samples support this theory? Horvath suggests that one response to cancer – and probably other abnormal perturbations of the cellular machinery – is to ratchet up maintenance of the epigenome, presumably to try to bring aberrant genes back into line through strong epigenetic control. When the tumour-suppressor genes that would trigger this response are themselves broken, little is done to stimulate extra epigenetic maintenance, and the clock continues to tick much as it does in the non-cancerous cells of the body.

Horvath's novel epigenetic clock is obviously a great new tool in the hands of doctors and medical researchers, who can use it to screen for tissues and organs that are showing signs of accelerated ageing – possible evidence of cancer or some other disease that warrants investigation, such as liver damage from heavy drinking, for example. Indeed, Janet

Lord and her colleagues in Birmingham are using Horvath's clock to investigate what's going on in victims of trauma, be it physical or psychological, who often remain weakened by their experience and more vulnerable to early death. Has the effort to recover from trauma speeded up their biological clocks, the investigators are asking? Are these people ageing faster than they should be? 'I don't know the answer yet,' says Lord. 'Basically we're still analysing the data. But if [our hunch] is correct, you might say why would we want to know? Well, you can then do something about it. With all these anti-ageing drugs coming along – we can try some of these out on trauma victims to see if we can give them a longer lifespan.' Her argument is valid: more recent research by Horvath and others suggests that an acceleration of the biological clock does indeed add to the risk of premature death.

Far more fundamental questions are also being asked of the epigenetic clock. Is it, for example, a passive reflection of the ageing process being driven by other forces, or is the epigenome *itself* a driver? In other words, can we do anything to reverse ageing by artificially manipulating the natural switches on our genes? 'I think that's super-interesting from a scientific perspective,' says Wolf Reik, who heads the epigenetics programme at the Babraham Institute, in Cambridge, UK. 'Is it a downstream thing? Is it an upstream thing? That, I think, is very important.'

To answer such questions we'd need to be able to tweak the epigenetic mechanisms and see whether this has any effect on the rate of ageing. 'Such an experiment would be impossible in humans, for both ethical and practical reasons,' says Reik. So he and his colleagues at Cambridge, looking for a model organism to work with, have developed an epigenetic clock for mice – one that, like Horvath's, is based on changes over time in the methylation patterns of

DNA, but that uses 329 different reference sites on the mouse genome from those used in the human clock.

The Cambridge team validated their clock by showing that it ticked faster when they made lifestyle changes to the mouse models that are known to shorten the creatures' lifespan, such as feeding them a high-fat diet and interfering with female hormones by removing their ovaries. Conversely, they found that the clock ticked more slowly in mice genetically programmed for dwarfism – a condition known to increase the lifespan of these creatures. Now Reik and his partners are busy investigating exactly how the epigenome responds to such lifestyle-related cues, and whether they can find small molecules that can mimic the effects, or develop tools to edit the epigenome directly. 'This should reveal whether ageing is directly influenced by DNA methylation patterns, or if ageing is a read-out of a story already written in our genomes,' say Reik and his colleague, Oliver Stegle, in an article about the group's work.

Whatever the answer to that, Reik believes the possibility of rewinding the clock of ageing is real. 'Oh definitely,' he says. Scientists are already able to generate 'induced pluripotent stem cells' (iPSCs) – that is, stem cells with the potential to become virtually any kind of specialised cell – from existing adult cells. 'That's a good starting point,' he says. Those experiments show that 'the potential for reversing ageing is absolutely there'.

CHAPTER FOURTEEN

Stem cells – back to fundamentals

Induced pluripotent stem cells are the brainchild of Japanese biologist Shinya Yamanaka, whose revolutionary work with the raw materials of life won him a Nobel Prize in 2012. Yamanaka was born in 1962 in Osaka where his father, an engineer, ran his own small company designing and building machinery parts. Shinya remembers as a child being curious about how things work and dismantling clocks and radios, which he rarely managed to put back together again. He remembers, too, getting in a row with his mother when he set fire to a quilt in the family home while playing with the experimental kit that came with his monthly science magazine for elementary-school kids. Yamanaka was a slight child – 'skinny', according to his father – so he took up judo with determination. But he quit after serious injury a few years later, and turned instead to his music, playing guitar and singing in the folk band he had started with some classmates.

Academically, maths and physics were his strengths at school. Yamanaka chose to go into medicine rather than following his father's path into engineering, and eventually qualified as an orthopaedic surgeon. But he found surgery more difficult to master than he had anticipated and, as confidence in his abilities waned, he became acutely aware of how often medicine has no cure for a patient's condition, no matter how skilled the practitioner. 'Painful and unforgettable bedside experiences finally drove me to switch my goal from becoming a surgeon who would help free patients from pain to becoming a basic scientist who would eradicate those intractable diseases by finding out their mechanisms and ultimately a way of curing them,' he told his audience at the Nobel Prize award ceremony.

Yamanaka spent some years in the US learning the ropes in molecular biology before returning to Japan and beginning to focus on stem cells, whose properties had begun to fascinate him when he was creating genetically modified mice. Right up to the mid-twentieth century the assumption had been that once a stem cell has set off on the path to becoming a mature adult cell – a specialist liver, heart, brain, blood cell, whatever – there is no going back; differentiation runs in only one direction. Repair and maintenance of tissues draws on the pool of immature stem cells, with each tissue type essentially having its own pool of semi-committed stem cells waiting in the toolbox. The only stem cells capable of becoming virtually anything – that is, 'pluripotent' – are embryonic stem cells, and the more scientists learnt about them, the more excited they became at the obvious potential of these cells to create novel treatments for disease. But since the beginning, the idea of harvesting cells from human embryos has met with huge public and political resistance. Scientists were soon thrown back on their own ingenuity.

In December 1999, Yamanaka acquired his own lab for the first time, and he set himself and his team the goal of generating stem cells with limitless potential using adult cells rather than embryonic ones – in other words, learning how to run development in reverse, right back to point zero. 'However,' he told his Nobel audience, 'I knew that making pluripotent cells from somatic cells would be extremely difficult, and when I started this project with my lab members at NAIST [the Nara Institute of Science and Technology in Ikoma, Japan], I was not sure if the goal could be achieved in my lifetime.'

The long-held conviction that differentiation runs in only one direction had already been overturned in 1962 – by a nice touch of coincidence, the year of Yamanaka's birth – by John Gurdon, a biologist at Oxford University, UK, with whom the Japanese scientist shared, so many years later, the 2012 Nobel Prize. Working with frogs, Gurdon had removed the nucleus (where the DNA is housed) from an egg cell and replaced it with a nucleus taken from a mature gut cell, and the egg had gone on to develop into a tadpole, just as an egg cell should. This was cloning – the same basic technique as was used to create Dolly the Sheep in Scotland in 1996 – and it showed that adult DNA can be stripped of its specialist instructions and reprogrammed.

Then came further evidence that the direction of travel in cells can be reversed when Takashi Tada at Kyoto University reported in 2001 that he had managed to reprogramme mouse thymus cells by fusing them with embryonic stem cells. Yamanaka, taking a step back, figured that such stem cells must contain some factors that maintain their pluripotency, since every fertilised egg starts life with its epigenetic clock reset to zero – that is, with very little epigenetic 'memory' carried over from the parents' DNA*. Perhaps these factors alone could be introduced into a cell, without the need for the infinitely fiddly and uncertain job of transferring whole nuclei. He set his lab the task of finding out what the factors might be – and, essentially, which genes are responsible for producing them. Other labs were looking for the same things, and by 2004 they had together identified 24 candidate genes that seemed to be involved in pluripotency.

* I say 'very little epigenetic memory' because the discovery of transient environmental influences that are carried over from one generation to the next is a fascinating new frontier in epigenetics/genetics.

The following year, 2005, Yamanaka's team managed to produce embryonic-like stem cells from adult cells in mice using just four of the genes, which they introduced into the DNA of the adult cells using specially tailored viruses as vectors. These four genes, they noted, had to be highly active in the cell to induce pluripotency – and presumably worked by removing the epigenetic switches that had decided its fate as an adult cell. But the experiment was so much simpler than Yamanaka had expected that he was inclined to distrust the results.

Fellow scientists were similarly sceptical when he presented his team's findings at an international stem-cell conference in Toronto, Canada, in June 2006. Mindful of a recent scandal when a Korean scientist's claim to have created human embryonic stem cells by cloning turned out to be fraudulent, Yamanaka got his researchers to repeat the experiments over and over again. Then, confident of their results, they published them in the journal *Cell* in November 2006, labelling their new creations 'induced pluripotent stem cells', or iPSCs.

Anxious to try out in human cells what he and his team had learnt from their mouse experiments, Yamanaka had moved in 2004 to the Institute for Frontier Medical Sciences at Kyoto University, the only place in Japan licensed to do such work. The year after their paper in *Cell*, his lab managed to produce human iPSCs – using the same technique as with the mice – from skin cells taken from the face of a 36-year-old woman and connective tissue cells from a 69-year-old man. What's more, his team managed to *re*-programme these artificially created stem cells into neurons and heart-muscle cells (which even began to pulse!) – a good sign that reversing the clock had not destroyed their potential to become something new. But his team had no time to rest on their laurels: getting wind of the fact that two other groups were hot on their heels, Yamanaka's lab worked like demons to get there first. In the event he and his colleagues published their findings in *Cell* in November 2007, just weeks ahead of the competition.

Today the 'Yamanaka factors', as the four genes whose activity is responsible for reprogramming are now commonly known, are used routinely with ever more refined techniques to produce iPSCs for research – if not yet for producing the replacement tissues and organs and spare body parts for real patients that seemed so tantalisingly close in 2007. Unsurprisingly, there have been – and still are – many wrinkles to iron out. For example, one of the factors, a gene known as Myc, has the propensity to trigger cancer. Another challenge is the fact that iPSCs do seem to retain some shadowy memory of their former lives as adult cells – thought to be due to some epigenetic tags that are extra hard to scrub off. Nevertheless, it wasn't long before scientists began to look beyond the cultures in their lab dishes to wider horizons. What would happen if they tried using the Yamanaka factors to turn back the developmental clock in whole organisms?

A number of labs tried it with mice, but a couple of groups in Spain and Japan who published their experiments in 2013 and 2014 reported dire consequences. The mice did not survive long, dying either from cancer after developing multiple tumours in cells that had lost their controls, or from organ failure as dedifferentiated cells lost their identity. Then in 2016, scientists at the Salk Institute in La Jolla, California, reported spectacular results from their experiments with mice that hit the media headlines worldwide.

The group was led by Juan Carlos Izpisua Belmonte, a smiling, soft-spoken Spaniard now in his late fifties, with a reputation for 'fearless' risk-taking and pushing at the boundaries of science – and often medical ethics, too – in his quest to understand the workings of nature; most especially how bodies of all sorts grow, develop and repair themselves. He is, for example, associated with such controversial issues as three-parent embryos and attempts to grow human organs in pigs. Izpisua Belmonte, who has been a professor at the Salk since 1993 and helped set up the Center of Regenerative Medicine in Barcelona, Spain, in 2004 – is also renowned for his amazing intellect and strong work ethic. His schedule, he told an interviewer for *STAT News*, is 'science 25 hours a day'.

Izpisua Belmonte was born into a poor farming family in rural Spain and raised by parents with very little education. He himself had to leave school aged eight to work in the fields, but managed to return to school at 16 and make it to university, and beyond to a PhD. His introduction to the mysteries of the developing embryo came when he joined a lab in Heidelberg, Germany, and the way stem cells work to generate the almost infinite variety of life forms has fascinated him ever since.

His lab was quick to start experimenting with iPSCs, and has used the Yamanaka factors to reprogramme a number of different cell types – a process Izpisua Belmonte refers to more specifically as 'global epigenetic remodelling'. His experiments have included human cells from centenarians and from patients with Hutchinson–Gilford progeria syndrome (HGPS), a disease that affects children, leading to rapid development of some (but not all) of the features of old age. In both, the reprogramming process succeeded in resetting the length of telomeres, the expression of genes and the levels of oxidative stress.

Turning from their Petri dishes to living creatures, Izpisua Belmonte's lab created genetically engineered mice that mimic HGPS in humans, developing some of the pathologies of regular ageing but on a speeded-up timescale that would give the researchers a quick answer to whether or not it is possible to rejuvenate whole organisms by reprogramming. Mindful of the dreadful fate of the mice in previous experiments, and working on the evidence that reprogramming is a stepwise process, the team decided to activate the Yamanaka factors in their mice intermittently and in short bursts by adding a special drug to their drinking water. This way, they figured, they could control how far back in time they took the creatures' cells and thus avoid 'dissolving' their organs by completely obliterating the epigenetic memory of what they were supposed to be, or triggering the wild, ungoverned growth of cancer.

They figured right: their strategy of partial reprogramming succeeded in slowing down the ageing process in a host of tissues and organs, including the skin, kidneys, stomach and muscles of treated progeria mice. What's more, the animals lived on average 30 per cent longer than the untreated controls. 'The mice that were treated with these factors had tissues that were better looking, they were more healthy, and they didn't accumulate the ageing hallmarks,' said Pradeep Reddy, a member of the research team, when their results were published. 'Together, all these helped them in the prolongation of their lifespan.'

The scientists also wanted to know whether they could use the strategy to reverse the age-associated decline in normal, but already elderly, mice. They tried it and it worked, restoring, among other things, the depleted supplies of healthy, functioning stem cells and the capacity of the pancreas (so important in diabetes) and muscles to repair themselves after injury. 'It's pretty amazing, if you think about it!' commented Wolf Reik. 'It's a very interesting strategy. It clearly speaks to the fact that [the pluripotency genes] can change the whole thing. And to the fact that if you can do a little bit of iPSC, you may be in a sweet spot there.' It also adds strength to the theory that epigenetic mechanisms are among the active drivers of ageing, not just passive reflectors of the process.

'Obviously, mice are not humans and we know it will be much more complex to rejuvenate a person,' commented Izpisua Belmonte when the Salk announced the results of his team's work. 'But this study shows that ageing is a very dynamic and plastic process, and therefore will be more amenable to therapeutic interventions than what we previously thought.'

Now scientists are looking for chemical compounds – drugs – that can do the same job as the Yamanaka factors, and that can be used intermittently to turn the rejuvenation process in their lab mice on and off at will. 'I think that will be much safer, much more secure, and could allow us to investigate whether that process can be applied in humans,' said Izpisua Belmonte.

iPSC technology holds promise for other anti-ageing treatments, too. As already mentioned, virtually all our tissues have a supply of bespoke stem cells to draw on for maintenance and repair. But for a number of reasons these stem cells get less and less efficient as we age. They tend to accumulate mutations in their DNA with repeated rounds of division. Their epigenomes get burdened with extra tags and lose their efficiency at directing gene expression. 'And we know part of the reason [they get less efficient] is because senescent cells make secretions that prevent stem cells from proliferating and then differentiating,' says the Buck Institute's Judith Campisi, whom we met in Chapter 4. 'In that case getting rid of senescent cells might then help rejuvenate the tissue. But there are other scenarios where you do literally run out of stem cells, and then getting rid of senescent cells won't do anything.'

This depletion of the raw materials for repair is known as 'stem cell exhaustion'. It's one of the classic hallmarks of ageing and a particularly striking example is provided by the case of Hendrikje van Andel-Schipper, a Dutchwoman who died in 2005 at the age of 115. Before her death, Hendrikje van Andel-Schipper was considered to be the oldest person in the world and, apart from being a little frail, she was in remarkably good health; she still had an unclouded mind, an interest in current affairs and a wealth of vivid memories gathered across the centuries. So what was the secret to her long and healthy life? To the delight of researchers, Van Andel-Schipper had agreed to leave her body to science. Henne Holstege, a geneticist at the VU University Medical Center in Amsterdam, headed the team that analysed her blood, and they were in for a surprise: the great majority of Van Andel-Schipper's white blood cells, they discovered, had originated from just two specialist stem cells (known as haematopoietic stem cells, which give rise to crucial immune cells).

At birth, we humans have around 10,000–20,000 of these white blood specialists, packed away mostly in our bone marrow, and every day around 1,300 of them are actively involved in topping up the system. But close analysis of Van Andel-Schipper's blood DNA indicated that at the time of her death her pool of haematopoietic stem cells had been virtually drained. 'The way the mutations were distributed over all the blood cells, it could only mean they were parented by two blood stem cells,' Holstege told *The Scientist.* She and her colleagues found a clue to the depletion of the haematopoietic stem-cell population in the extremely short telomeres in Van Andel-Schipper's blood compared with that of other tissues. This indicated that they had neared the end of their replicative lives – and presumably the other 19,998 had already done so.

When you run out of maintenance materials through stem-cell exhaustion, says Campisi, the only solution is to create new stem cells, and there the promise is transplantation. 'We take a little skin biopsy; we make your own cells pluripotent, so there's no immune issues, right? And then we redifferentiate them into neural stem cells, or muscle stem cells, for example, and implant them.' A big challenge here, she says, is delivery. 'If we want to just get some stem cells to make new cartilage in your knee, that's pretty easy. But if you're running out of stem cells in all of your long muscle, how are we going to get all those cells [to where they need to be]?' Some stem-cell populations are better than others at homing in on areas that need regeneration, but no one yet knows why, nor what they can do to energise the populations that seem reluctant to migrate. And hanging over whatever strategies they devise to use stem cells in anti-ageing therapy, warns Campisi, is the shadow of the Big C. 'Every time a cell divides there's a risk that it will become cancerous. That's going to be a main concern.'

The regenerative capacity of our bodies does not, however, depend only on the quantity and quality of our stem cells. It depends also on the condition of the blood that supplies these cells with growth factors and other vital chemicals that drive their activity. How do we know? What follows is not for the squeamish, for it involves the joining together of live mice down the flanks so that they share their circulation, a mingling of the blood. This is parabiosis.

CHAPTER FIFTEEN

Something in the blood?

Parabiosis, derived from the Greek meaning 'living alongside', was pioneered in 1864 by French physiologist and politician Paul Bert, who was simply curious to see whether animals stitched together through the skin would eventually come to share one circulatory system. He tried it with his lab rats, joining them along the flank, and it worked, earning Bert a prize for experimental physiology in 1866 from the French Academy of Sciences. This was proof of principle, but the scientific community showed little interest in parabiosis as a research tool until early the next century, when what amounts to whole-animal grafting began to be used on a variety of creatures besides rodents, including frogs and insects, to study all kinds of biological phenomena, both healthy and diseased.

The first person to think of grafting animals together to study ageing, in 1956, was Clive McCay – the animal-husbandry scientist we met in Chapter 8 experimenting with caloric restriction for the American cattle industry. McCay, as you will recall, was mighty preoccupied with ageing. Keen to find out whether young blood might be a fountain of youth, he stitched together 69 pairs of rats in various combinations of ages. But his techniques were crude and led to some gruesome deaths from aggression between conjoined creatures, survival of one creature at the expense of the other, or from parabiotic disease, a mysterious condition thought to be an immune reaction that sets in at the point where the joint vasculature is developing.

In those creatures that survived the procedure, however, McCay did find evidence of a rejuvenating effect on the tissues of the older partners. Their bone density, for example, improved significantly. And when he joined a mouse that had been subjected to caloric restriction to one that had fed at will, he saw some evidence of life extension. But his data, though intriguing, were extremely limited and mostly anecdotal. It wasn't until the 1970s that researchers, also pairing old mice with much younger ones, obtained strong evidence of extended lifespan. The older partners in their pairings lived some four or five months longer than controls.

The next big milestone in the use of parabiosis* for ageing research came in 1999, when stem-cell biologist Amy Wagers, then at Stanford University in California, was looking for ways to study the fate of bone-marrow stem cells circulating in the blood. At the time, many people held the erroneous belief that these bone-marrow cells were almost as potent as embryonic stem cells in what they could become. Wagers was working as a postdoc in the lab of Irving Weissman, who had been using parabiosis for many years to study regeneration in sea squirts, small invertebrates that live on the ocean floor. Weissman recommended that

---

* Today, everything is done to make parabiosis as pain- and stress-free as practical for the animals being joined together. The experiments described here will have been carried out according to the US National Institutes of Health Guide for the Care and Use of Laboratory Animals. The mice will typically: have shared a cage for at least two weeks prior to the operation to ensure compatibility; been operated on under sterile surgical conditions and anaesthesia; been kept warm by a heating pad during recovery; been given potent pain relievers for the surgery and until they heal afterwards, as a matter of course; been observed daily for signs of distress that required further attention; had their food and water placed within easy reach to minimise physical effort and discomfort during the recovery period; experienced a similar procedure during subsequent reversal of the operation.

his postdoc use the procedure for her research. Wagers did so, thereby proving conclusively that bone-marrow stem cells, which sustain the immune system, are not able to generate other specialist cells – and certainly not neurons in the brain, which some had suggested. Her experiments were an inspiration to others in the field of stem-cell biology – prominent among them, Irina and Mike Conboy, cell biologists who work together today in the bioengineering department at the University of California, Berkeley.

On a hot Sunday morning in August 2016, I visited the Conboys in their office on campus. Irina had just come in from a swim, still towelling her hair. Mike, in a casual check shirt, slacks and sandals, joined us a while later as Irina was describing to me how she had left her native Russia to study in the US, and how she had become involved in ageing research.

The subject has preoccupied Irina from a very young age, she told me. 'I remember the exact moment very vividly, as if it was yesterday.' She was six years old and, watching her grandmother, she noticed the wrinkled skin on the old lady's arms, so very different from her own firm young flesh. 'Her arms just looked very, very old,' Irina said, 'and I realised that this thing will happen to me at some point, because I am growing taller, and I'm changing, and the natural progression of that is to become like my grandmother.'

Recognising that the old lady was also moving towards death sharpened the focus. 'I became very sad and I wanted to help my grandmother not to die, and of course my parents.' So the urge to understand and conquer ageing joined the other obsessions in Irina's young life 'like becoming a princess and a gymnast and a gold medallist … all kinds of stuff!' she laughed.

After college in Russia, Irina moved to the US in the early nineties, looking for better education opportunities. She did a PhD at Stanford University and then began focusing on muscle stem cells, looking particularly at why they become sluggish with age. She had intended eventually to return to Russia, but everything changed when she met and married fellow scientist Mike soon after arriving in the US. Together, the two have made a name for themselves worldwide with their studies of blood, and the keys it holds to some of the mysteries of ageing.

Irina's postdoc work at Stanford was in the lab of Tom Rando, where Mike was also a postdoc. As part of their research, they took muscle stem cells from old and young mice and put them in culture with the serum (the liquid component of blood) of old and young mice in various combinations. The results were intriguing: the researchers found that serum from old mice seemed to inhibit the activity of stem cells from young mice, while young serum stimulated the activity of stem cells from old mice. When the scientists bathed stem cells in a 50/50 mixture of serum from old and young animals, they found that the old serum dominated the picture, inhibiting activity in stem cells from whatever source. So how does this play out in living organisms, they wondered?

The Conboys were wrestling with other questions, too. Why, they wondered, does ageing seem to be a generalised phenomenon, with so many tissues in our bodies declining at the same time? Irina already had her own unorthodox theory about our failing powers of regeneration with age. It is not the stem cells that age, she suggested, but the environment in which the cells live that becomes depleted so that they no longer get the stimulation they need to do their job. Could there be something in the circulation that coordinates their activity?

The story goes that in 2002, Irina was presenting Amy Wagers's paper on her work with bone-marrow stem cells at one of the journal-club gatherings of her department. When she began describing Wagers's experiments with parabiosis in which pairs of young mice shared their blood supply, Mike, who had been sitting quietly at the back of the gathering, suddenly recognised an opportunity. Cornering his wife and their lab chief Tom Rando after the meeting, he asked: why don't we use the same system, but connect young mice to old ones? 'This had never been done before for stem-cell biology or regenerative medicine,' explained Irina, and she and Rando saw immediately the attraction of such experiments.

They enlisted the help of Amy Wagers, who joined up most of the animals in the Rando lab's initial experiments, teaching Mike Conboy the technique along the way so that he could carry out the procedure in future. Within a week or two their paired mice – combinations of young/old as well as same-age controls – were sharing their circulation. After five weeks the research team euthanised the animals in order to analyse tissues from the brain, muscles and liver. These organs represented the three germ layers of cells – ectoderm, mesoderm and endoderm – that are formed in the early embryo and give rise to every organ in the body.

'For all of these tissues there was profound rejuvenation of stem cells and regenerative capacity in the old mouse,' said Irina. 'And there was noticeable decline in the young mouse.' The team published their results in the journal *Nature* in 2005. 'But because everybody was so excited about the rejuvenation part and not so excited about the premature ageing of the young partner,' commented Irina, 'the paper – kind of in the title and the description – was I think a little bit skewed towards rejuvenation.' The researchers used some clever genetic sleuthing to make sure the regeneration of tissue was the work of the resident stem cells and not ones that had migrated from the young mouse in the bloodstream.

But why, in their live experiments, was the rejuvenation of the old mouse so much more pronounced than the premature ageing of the young one, when their cell cultures in the lab had led them to expect the exact opposite? Because, said Irina, it's not only their circulation that conjoined animals share. 'You have the entire young animal, not just a bag of blood.' The old mouse benefits from the activity of the robust young liver and kidneys that can remove all the garbage of ageing from the circulation, and a lot more besides. 'Even blood pressure becomes better. And there's better oxygenation because of the young lungs, and better insulin/glucose balance,' she says. 'By contrast, now the young animal [suffers because it] has to maintain a pathological body with inflammation.'

Among the garbage circulating in old blood are inflammatory molecules released by senescent cells, as we've already seen in previous chapters. But following their parabiosis experiments, the Conboys soon identified another factor in old blood that causes harm. This is TGF-beta (transforming growth factor), a molecule which is overproduced as mice (and humans) get older. What's more, the cells that receive signals from TGF-beta sprout more receptors – or docking sites – for this messenger molecule, thus compounding the effect of overproduction, which is to inhibit regeneration, particularly of muscle and brain cells.

The researchers also identified one of the factors circulating in *young* blood that helps to rejuvenate old tissues. This is oxytocin, a hormone produced in the brain and familiar to many people as the substance that aids contraction of the uterus in childbirth. 'Oxytocin has a direct effect on muscle stem cells,' explains Irina. 'It has receptors on muscle stem cells, and without oxytocin muscle doesn't repair well; it does not function well, and it is replaced by fat. With ageing, oxytocin levels in blood decline about threefold and oxytocin receptors become less present.'

The Conboys have found that by normalising levels of TGF-beta in elderly mice – or by adding other growth factors, or the beneficial oxytocin, to their blood – they can 'reset' the system so that 'stem cells wake up, start dividing, and doing their regenerative thing,' said Mike, just as Irina's unorthodox hypothesis had suggested. 'If what we can do with a mouse translates well to humans, it seems quite encouraging – this strategy could be very valuable for people, as an adjunct to surgery or after an accident or something, a trauma.'

TGF-beta inhibitors, oxytocin and a variety of growth factors are already in the medicine cabinet to treat a variety of conditions, but translation of their work for clinical use is not the Conboys' main preoccupation, which is to tease out what's happening silently, ceaselessly and unseen deep within our bodies. 'I mean, we do the work and we publish it and hope someone else kind of reads it and maybe will try something like that for a therapy,' said Mike.

Enter Tony Wyss-Coray, a neuroscientist whose lab is on the next floor down from Tom Rando's at Stanford University. Looking at brain tissue from their conjoined animals, the Rando team had found impressive regeneration of neurons in the hippocampus – the region where memories are stored – in old mice, and shrinkage of this part of the brain in young mice. They had, however, left out these findings from their original 2005 paper to avoid delays in publishing after reviewers had asked for further work on the brain tissue. But Wyss-Coray picked it up and ran with it, performing parabiosis studies of his own that confirmed the Rando team's results and convinced him they were on to something big. He went on to find that plasma alone had the same effect, perking up the brains of befuddled elderly mice so that they could lay down new memories, learning to find the only hole in a field of holes on a big board that offered escape from a scary flashing light.

This would be the human equivalent, he told his audience at a TED talk in June 2015, of 'finding your car in a parking lot after a busy day of shopping' – an everyday feat that's distressingly difficult for people who are losing their minds. And a research result that clearly gave the scientist huge optimism for the future. 'Now, I don't think we will live forever,' he told his audience. 'But maybe we discovered that the fountain of youth is actually within us, and it has just dried out … If we can turn it back on a little bit, maybe we can find the factors that are mediating these effects; we can produce these factors synthetically and we can treat diseases of ageing, such as Alzheimer's disease or other dementias.'

To test his theory that we, too, have factors in our blood that have powers of rejuvenation, Wyss-Coray used plasma from young human blood to treat the elderly mice in his experiments. And so encouraged was he by the results that, by the time of his TED talk, he had already, the previous year, set up his own small company, Alkahest, in San Carlos, California – with seed funding from a rich family in Hong Kong who had a history of Alzheimer's disease – to run a small-scale human trial.

The trial, led by Stanford neurologist Sharon Sha, involved 18 people aged between 54 and 85 with mild to moderate Alzheimer's disease. Once a week for four weeks they received injections of plasma from volunteer blood donors aged between 18 and 30 years, or else a saline solution as a placebo. At the end of the treatment their brains were scanned, they were given cognitive tests and their caregivers were asked to assess whether or not they had seen improvements in performing simple tasks of daily living, such as getting dressed, preparing a meal or shopping. The results, announced in November 2016, were a big disappointment. Several participants dropped out of the trial before the end; no improvement was found in the cognitive function of those who remained; and only mild improvements in performing everyday tasks were reported by caregivers.

Irina Conboy was dismissive. 'The scientific basis for the trial is simply not there,' she told *Nature News* when the results came out. No one yet knows the cellular mechanisms targeted by the good stuff in blood plasma. Furthermore, experiments like Wyss-Coray's, with the mice, the flashing light and the maze of holes designed to test the effects of young blood on cognition, had not been replicated independently, she commented. 'And there has never been a test with a mouse model of Alzheimer's.'

Intrigued by the questions raised by their original parabiosis studies, the Conboys had been doing further research of their own, which they published the same month Wyss-Coray released the results of his human trials – and which perhaps explains Irina's somewhat exasperated response to his announcement. In particular, the two scientists were keen to establish just how much of the effects they had seen with parabiosis could be attributed to factors in the circulating blood, rather than to the shared activity of the animals' organ systems.

They devised a contraption with computer-controlled pumps that allowed for the exchange of precisely measured quantities of blood between paired animals that were not stitched together and thus did not share organ activity. Once again, they paired a range of different young/old mice, with pairs of same-aged animals as controls, and looked at the effects on the same tissues as before – brain, muscle and liver – of sharing blood in equal parts between two animals.

What they found was fascinating. As they had seen in parabiosis, young blood helped old mice repair muscle tissue that had been damaged, while old blood considerably weakened the muscles of young mice. Again, old liver was somewhat rejuvenated by young blood, while young liver was prematurely aged by old blood. But the biggest surprise was the effect seen in the brain. 'Under no circumstances did young blood improve brain neurogenesis in our experiments,' said Conboy. 'Old blood appears to have inhibitors of brain-cell health and growth, which we need to identify and remove if we want to improve memory.'

The overall picture from this new experiment was much closer to what they had seen in their lab dishes before they started parabiosis – namely, that the effect of old blood is stronger at suppressing young cells than young blood is at rejuvenating old cells. 'Our study suggests that young blood by itself will not work as effective medicine,' said Irina Conboy in the press release that announced their results. 'It's more accurate to say that there are inhibitors in old blood that we need to target to reverse ageing.'

Under the circumstances, it looks as if Wyss-Coray's trial of young plasma as a treatment for Alzheimer's was premature and destined to disappoint – another in a long line of disappointments with this most complex and still somewhat mysterious disease of the brain.

CHAPTER SIXTEEN

The broken brain

On 25th November 1901 in Frankfurt, Germany, a distraught husband took his wife to see a psychiatrist because of her increasingly bizarre behaviour. Fifty-one-year-old Auguste Deter had become delusional; she was blindly jealous of her husband, and so sure that someone was out to kill her that she would have fits of fearful screaming. She suffered hallucinations and loss of memory, and was disorientated even in her own home. Deter was admitted to the Hospital for the Mentally Ill in Frankfurt under the care of one Alois Alzheimer, the psychiatrist to whom she had been referred by her husband, and who was so intrigued by her striking cluster of symptoms that he continued to follow her case until her death nearly five years later. By then, Alzheimer had been transferred to a clinical practice in Munich, but after a post-mortem of Deter's body he was sent her brain for examination by the director of the Frankfurt institution that had cared for her.

Under the microscope Alzheimer saw, and described fully for the first time, hard clumps of protein, clustered barnacle-like around the bodies of nerve cells. And within the neurons themselves, the microtubules that form the skeleton and communication network of the nerve cell and its branches were tangled and collapsed. These were the 'plaques and tangles' that are today the hallmarks of the disease that bears his name. Of the many different forms of dementia – the catch-all label we give to the most extreme examples of cognitive impairment – Alzheimer's disease is the most common, especially in people over the age of 65, accounting for around 75 per cent of all cases. Alzheimer himself, however, dubbed the disease he'd described as 'pre-senile dementia', because of the relatively young age of his first case. Neither he nor anyone else in the mental-health community equated it with what they were seeing in older people; 'senile dementia' was put down to hardening of the arteries, and widely accepted – with what seems now like an extraordinary lack of curiosity – as part of the normal process of ageing.

For half a century Alzheimer's disease was considered a rare curiosity that most knew from their medical textbooks but few expected to see in their clinics, and even fewer were interested in investigating. The main purpose of those who were doing further research was simply to distinguish it from other forms of dementia that affect younger adults.

Then in 1968 came a paper from three British scientists that challenged the shaky assumptions about dementia in old age. At the time, Freudian ideas were popular and mental disorders were frequently ascribed to the psychological effects of adverse experiences in childhood. But Martin Roth, Professor of Psychiatry at the University of Newcastle upon Tyne, swam against the tide: he was interested in exploring the biological underpinnings and causes of psychiatric disorders. Roth persuaded his colleague, the Professor of Pathology at Newcastle, Bernard Tomlinson, to look at the brains of old people who had died of dementia. Together they recruited the help of Gary Blessed, a senior doctor on the neurology unit, to carry out clinical assessments of elderly patients on the long-stay wards of the local mental institution, as well as elderly patients without dementia seen at the general hospital, as controls. They asked Blessed also to obtain permission for post-mortems of both sets of patients so that Tomlinson could study their brains.

For that initial research project, Tomlinson looked at the brains of 78 people, most in their late seventies. Contrary to expectations, he found that of those who had suffered dementia, the majority had the gummy plaques and tangles described by Alzheimer more than 50 years earlier. He noted, too, that the amount of damage he found in a person's brain correlated with the degree of dementia described by Roth and Blessed.

Tomlinson – dignified, old-school, but open-minded and with a sense of fun, according to colleagues sharing their memories of the man for his obituary (he died in May 2017 aged 96) – is honoured in some quarters as 'the father of neuropathology' for his groundbreaking work on Alzheimer's. But long-held beliefs are not easy to overturn, and it took a while for the three Newcastle scientists' findings about dementia in elderly people to be widely accepted, and for the implications to sink in: here were the signs not of some inexorable process of degeneration, but of a thoroughgoing disease; a disease, what's more, that already affected millions of people worldwide, and that threatened many millions more. Clearly Alzheimer's warranted serious attention, and since then a huge amount of research effort has gone into teasing out what's happening in a demented brain and how and why the disease develops.

The first wave of Alzheimer's research looked for evidence of faulty signalling between brain cells in the hope of developing drugs that could make up for lost capacity, and here I shall take a short diversion to set the scene. In the years before Tomlinson and Co. made their discovery about the pervasiveness of Alzheimer's pathology, there had been an ongoing, spirited and often acrimonious debate among neuroscientists about how exactly the brain cells communicate. On the one side were 'the sparks' people, who believed signals were carried by electrical impulses. On the other side were 'the soup' people, who believed in chemical transmission of signals. Though the idea of chemical transmitters in the peripheral nervous system – that is, the network of nerves that branch out from the brain and spinal cord to the rest of the body – had been widely accepted by then, it was still fiercely contested as a mechanism in the central nervous system. Here, in the brain and spinal cord themselves, the sparks theory remained strong. Anyone who brought evidence of chemical transmitters in the body's command centre to neuroscience meetings was likely to be ridiculed, if they weren't simply ignored.

This was the experience of Arvid Carlsson, a Swedish neuroscientist who, in the late 1950s, discovered, serendipitously, the function of the neurotransmitter dopamine. Carlsson was running experiments on rabbits, and he found that suppressing this chemical compound in the brains of his laboratory animals induced Parkinson-like symptoms of muscle rigidity and tremors. He found also that by administering the drug L-dopa, a precursor of dopamine, he could relieve the rabbits' symptoms.

Carlsson was to be awarded a Nobel Prize in the year 2000 for his work on chemical signalling in the brain. But when in 1960 he took his dopamine data to an international symposium in London, he was met with 'near unanimous scepticism'. One of the delegates shook her head and told him his views 'would not have a long life'. And when the chairman, in his concluding remarks, said that nobody at the meeting had put forward any ideas about the possible action of chemical transmitters in the brain, the very point of Carlsson's presentation, 'the clear message to me was that I was nobody!' he told his audience at the Nobel Prize-giving in 2000.

Within five years, however, there had been a paradigm shift in attitudes as many young scientists got involved in research into neurotransmitters, and the evidence of their vital role in the central nervous system accumulated. The L-dopa story excited doctors, too, a small number of whom, in countries as diverse as Austria, Canada and Japan, began investigating its potential in their Parkinson's patients. By 1967 they had developed an effective dosage regimen, and L-dopa, which helps control tremors and relieve rigidity, became the mainstay of treatment for Parkinson's disease. It was the spectacular success of this new drug for a dementing disease that inspired scientists investigating Alzheimer's when research finally took off at about the same time.

This line of enquiry led, over the following decade, to the only two types of medicine currently in the cupboard for treating Alzheimer's, both of which target neurotransmitters. The first, and still most widely used therapy, works by boosting the levels of the signalling chemical acetylcholine to compensate for the progressive loss of the brain cells that use it to communicate, many of which are involved in forming memories. The second therapy does the exact opposite: it dampens down the effect of the neurotransmitter glutamate, which spills out in excessive quantities from damaged cells in the brains of people with Alzheimer's, and causes damage of its own at the same time.

But what of the plaques and tangles that are the hallmarks of Alzheimer's? First studied under electron microscopes in the 1960s, the plaques were seen to be close-packed bundles of fine protein fibres, and the tangles to consist mainly of paired, corkscrew-shaped filaments interspersed with some straight threads. It wasn't until the mid-1980s that the composition of these destructive lesions was discovered. The corkscrew filaments in the tangles were found to be made up of a protein called tau, first identified in 1975 as essential to the assembly and maintenance of the microtubules that form the skeleton and communications network within brain cells. The type of tau identified in the mid-1980s as the main component of tangles was clearly faulty, allowing the skeleton and the 'train tracks' of the communication network to collapse. Under a microscope, tau tangles 'look like hair clogging a drain', commented one observer.

The composition of the sticky plaques was discovered in 1984 by two scientists at the University of California, San Diego, George Glenner and Caine Wong, who identified it as a protein called beta-amyloid. But were these two proteins, tau and beta-amyloid – one, or the other, or both together – actually the cause of Alzheimer's disease, or just the consequence?

The scientists poring over this question had several clues that the amyloid at least was an active player. It was known that people with Down's syndrome have exceptionally high risk of developing dementia, often in early adulthood. The same year that Glenner and Wong identified the protein in Alzheimer's plaques as amyloid, they found the same protein in the brain of a Down's case. This was the first chemical evidence of a link between Down's and Alzheimer's, and one that led Glenner to speculate – wildly, because he had no way of proving it yet – that the problem lay with the gene that produces amyloid. If this were the case, the place to look for the gene would be on chromosome 21, of which people with Down's have an extra copy. The idea, wild as it was, caught the imagination of the Alzheimer's community, and the race was on to find the gene and be the first to publish.

Working backwards from the recipe of the protein revealed by Glenner and Wong, a number of labs managed three years later to clone the gene that encodes it – a gene known as APP (standing for amyloid precursor protein). And they found that it does indeed sit on chromosome 21. This was not yet a smoking gun for Alzheimer's disease, because there was no indication of how the regular gene might go wrong to start causing trouble, so no really strong evidence that APP was what they were looking for. But the spotlight remained tantalisingly on beta-amyloid, since the same year the APP gene was cloned, a group of doctors in the Netherlands described deposits of the same protein in the cerebral blood vessels of people with another, hereditary dementing disease that causes bleeding in the brain and stroke.*

Enter a cast of characters who have had enormous influence on Alzheimer's research – John Hardy, his colleague Martin Rossor and the Jennings family. Hardy, a genial, tousle-haired, stubble-cheeked man in his early sixties, trained as a neurochemist at Leeds University in the UK and spent his early years as a research scientist studying brain tissue from post-mortem cases for clues to what causes diseases such as Alzheimer's. Then in 1983, a paper came out in the journal *Nature* from a group at Harvard led by geneticist James Gusella. It described their discovery of the gene that causes Huntington's disease – the condition, incidentally, that took American folk singer Woody Guthrie to a mental institution aged 44 and finally killed him at the age of 55. 'For me that paper was a kind of "road to Damascus",' Hardy told me when I visited him in his lab at University College London. 'It made me feel that I should change fields; if I wanted to work out what causes disease, genetics was the way to go.' He decided to return from the US, where he had been working for the past 15 years, to a posting at St Mary's Hospital in London. There he began to learn molecular genetics and eventually started a research programme into the genetics of Alzheimer's.

Following Glenner and Wong's revelations about the composition of the sticky plaques, Hardy and his group had also joined the race to clone the putative Alzheimer's gene by the painstaking process of reverse-engineering of the

---

* It goes by the tongue-twisting name of Hereditary Cerebral Haemorrhage with Amyloidosis, Dutch Type, represented by the acronym HCHWA-D.

plaque protein to reveal its recipe. But they had spent fruitless and frustrating months following a false lead. In addition to the cloning exercise, however, they and others were casting their net wider in the search for the gene that might be involved in the disease by doing what are known as 'genetic linkage' studies, which home in on individual genes associated with a disease by first identifying their location on the chromosomes. In this case, the researchers were recruiting families with a history of early-onset Alzheimer's for the DNA analysis.

Hot on the heels of their success in finding the Huntington's gene, Gusella's lab had carried out a large linkage study in search of the Alzheimer's gene, and they too had narrowed the search down to chromosome 21, the one linked to Down's syndrome. But the Gusella group's linkage pointed to a stretch of the chromosome that didn't include the amyloid gene APP. This led everyone to assume there might be another Alzheimer's gene, or perhaps genes, on chromosomes, and threw them somewhat off the amyloid scent.

This is where Martin Rossor comes in. A neurologist working on the front line with patients as well as carrying out research, Rossor, a tall, slim man with a quiet manner, was engaged with John Hardy at St Mary's hospital in recruiting families with early-onset Alzheimer's for their linkage study. 'We advertised in the newsletter of the Alzheimer's Society, which had only very recently been formed, and we started collecting families,' he explained. 'We employed a nurse and we'd go out and examine people and take blood. My job as the clinician was to find families and make sure this was Alzheimer's disease, keep them on board. And John did the clever stuff in the lab.'

Hardy, Rossor and their team – a bunch of scientists with complementary skills – were guddling in the dark at first. The amyloid gene was only cloned the year they got the go-ahead for their linkage study, and for a while they turned up nothing exciting. At that time, remembers Rossor, linkage analysis 'was a phenomenally painstaking exercise'. Today you can sequence a gene in a matter of hours. But at that time, all you could do was chop up the genome and test every segment, getting smaller and smaller, until you found the bits – mutations in a single gene – associated with inheritance of the disease in question. 'It was months of work.'

The breakthrough came when Hardy spotted the 'fascinating' paper from the Dutch doctors about their distinctive stroke patients with the beta-amyloid clumps that clogged their cerebral blood vessels. He immediately got in touch with the group to ask if he could look at the DNA of the families in their report, and was soon on the plane to Leiden, Holland. There he met up with his Belgian collaborator, molecular biologist Christine Van Broeckhoven, and together they began collecting bloods from affected families in Antwerp, some 130km (80 miles) away. Linkage analysis focusing on chromosome 21 took them straight to the APP gene.

Competition was still extremely hot, and again Hardy's group found themselves racing against others on the same track. They sequenced their APP gene and found a mutation at exactly the same time as another group, based at New York University, announced similar success. The two groups of scientists published their findings back to back in the same edition of the journal *Science* in 1990.

This was the smoking gun for Alzheimer's and it gave Hardy, Rossor and their team a renewed sense of direction in their linkage studies back home. But they realised they needed to change their strategy. The London team had more than 15 families with early Alzheimer's on their books and, on the assumption that they all had the same genetic defect, the researchers had been pooling the bloods in their analysis. This was a mistake. They were to find out in due course that there are three different genes involved in familial Alzheimer's, and many more in the sporadic disease. Pooling all their families together created so much background noise that nothing leapt out at them from their comparative analyses. Realising that this might be why they were getting nowhere, they decided that henceforth they should analyse each family individually. What they needed to give them enough material to work on was a large family with a pattern of disease spreading across several generations, with uncles, aunts, cousins involved.

This is where Carol Jennings, her large family and a big stroke of luck come into the picture. By chance, I discovered that Carol's son John teaches at a school in Edinburgh, where I too live, and he agreed to meet up and tell me his family's story.

CHAPTER SEVENTEEN

Alzheimer's disease – the family that led the way

When it comes to human diseases, the role played by ordinary citizens in the advancement of science is rarely given due credit. It is often the desire for answers of the people most affected, their persistence and their willingness to get involved with research that leads to scientific breakthroughs. The Jennings family certainly played such a part.

John's mother Carol was an only child, but her father Walter was one of 15 children, only eight of whom survived to adulthood. Raised in a poor family where there were days when they didn't eat, Walter started work as a milkman at the age of 13, and worked his way up to become 'a credit agent for the local co-op in the days when they went door to door to collect payment for washing machines and things'. Walter was 'aspirational', says his grandson, and when Margaret Thatcher's government gave council tenants the right to buy their houses, he took the opportunity, and became the first home owner in his big family.

Then, in his mid-fifties, Walter started to have problems. His once meticulous record-keeping for the co-op became a little haphazard, and he sometimes forgot what day of the week it was. Out shopping with his wife, he would wander the supermarket aisles putting strange things in their trolley and sometimes in other people's baskets. 'There's a photo from my parents' wedding, which was in 1979,' said John, 'and there are stories about grandad losing his buttonhole and his gloves and that sort of thing. But it wasn't till a couple of years after that that he was diagnosed.'

John was born in 1985, and by the time he knew his grandfather the older man was no longer able to speak or walk unaided. By that time, too, four of Walter's brothers and sisters had also developed dementia. 'They all ended up in the same unit in this hospital in Nottingham,' says John. 'My great auntie was diagnosed, she was 48 I think, and her progression was very quick. She died by the time she was 55.'

Remembering that her paternal grandfather had had similar problems at a young age, John's mother Carol had become curious and had drawn up a family tree identifying others who had developed dementia. 'It must have felt a bit like a plague, I suppose, in the 1980s,' commented her son. 'It must genuinely have, for her to be surrounded by it in all her aunties and uncles.'

Her grandfather had fought in the Great War and his strange behaviour had always been put down to the effects of trauma or the poisoned gas attacks on soldiers in the trenches, and was not considered odd. But as she gathered the evidence of a much wider distribution of dementia among her relatives, Carol, a school teacher and a socially engaged kind of person, grew hungry for answers and believed someone would be interested in her family's case.

'Mum wrote lots and lots of letters,' said John. 'She wrote to the University of Nottingham because we lived nearby, and she also wrote to other universities and hospitals saying, "Are we of interest? We'll offer ourselves for research." But there was a widespread belief that there wasn't any genetic link, so she got a lot of refusals; and a lot of times people didn't respond.' John was lost in his own thoughts for a minute. 'I mean, how could anyone think there was *not* something strange going on?'

Then, in 1987, Carol heard of the research project at St Mary's in London, and wrote to Rossor and Hardy. This time she wasn't ignored or rebuffed. Her case history was exactly what the St Mary's team was looking for, and when, in 1990, they decided to analyse families separately, the Jennings family made the ideal study group. Carol reached out to as many family members as she could on her father's side – surviving uncles and aunts, cousins and second cousins – and managed to involve them in giving blood samples for analysis.

In 1991 Hardy, Rossor and the team struck lucky. The inconsistency in the DNA samples showed up on chromosome 21 and pointed to the APP gene. The scientists sequenced APP and found a mutation. 'We reported the finding on February 6th 1991 … There you are, that tells you how important it is that I can remember the date!' said Hardy with a smile. 'I knew it would be a life-changing event, and it has been.'

Their paper appeared in the journal *Nature*, with neuroscientist Alison Goate as the lead author. It became the most cited paper in the biomedical field that year, and led *Science Watch*, which analyses trends in research, to label APP work 'the hottest corner of biology'. (APP later got bogged down in a horribly tangled, acrimonious and funds-sapping dispute over intellectual property that saw the London team scatter and many of them, including Hardy and Goate, leave London for the US. But that's another story.)

Two other cases of familial Alzheimer's with the APP mutations were found that same year. Together these three discoveries formed the basis of what is known as the 'amyloid cascade hypothesis', which Hardy and his colleague, David Allsop, sat down and wrote 'without much thought', according to Hardy's candid memoir of those erratic, competitive years, so clear was the evidence all of a sudden. 'I had always thought of genetics as an independent way of testing hypotheses of causation,' he writes. 'There had been many competing theories for Alzheimer's disease, and I simply believed that genetics would allow a decision about these competing theories to be made. Genetic analysis told us that amyloid was the cause of Alzheimer's disease in these families and also in Down syndrome.'

The amyloid cascade hypothesis expounded by Hardy and Allsop remains the dominant theory of how Alzheimer's occurs to this day, and has had a huge influence on research. What it says, in essence, is that the accumulation of beta-amyloid in the cortex – the wrinkly outer layer of the brain that is the grey matter – is the trigger for the disease. It clumps together to form the plaques between the neurons, causing inflammation and subtly disrupting signalling as the glial cells – the 'scavengers' of the brain's immune system – become involved. The amyloid plaques drive the spread of the tau tangles – which may already be present in limited patches in the basal region, but doing little harm – throughout the brain in a characteristic pattern. The amyloid does this by attaching little epigenetic tags to the tau genes that make them produce an overabundance of the tau proteins. The gradual accumulation of these plaques and tangles in the brain is associated with increasingly widespread cell death and disruption of signalling, leading to dementia.

The amyloid cascade hypothesis was first published in 1992. That same year, scientists in the United States using linkage analysis with families affected by early-onset Alzheimer's homed in on chromosome 14 as another site of interest, and in 1995 the faulty gene on this chromosome was discovered. It was named presenilin 1, because, hot on its heels, a related gene was found to be mutated in familial Alzheimer's and named presenilin 2. This second gene, however, was on yet another chromosome – chromosome 1. Presenilins had been located through linkage analysis among members of a family known as the Volga Germans, whose ancestors had arrived in the US between 1870 and 1920 from two neighbouring villages on the west bank of the Volga River in Russia. Like the Jennings family, the Volga Germans were ideal candidates for linkage analysis since they shared common ancestry, which strongly suggested that an inherited genetic mutation was behind the high incidence of Alzheimer's among them.

Like APP, both the presenilins are involved in the production and accumulation of beta-amyloid, but at different stages from the precursor protein. When working normally, they are responsible for chopping up the amyloid precursor protein made by APP into a number of different functional forms.

Of the three mutant genes behind familial Alzheimer's disease, the one that was found first and that sparked the amyloid cascade hypothesis, APP, is by far the least common. However, it continues to blight the lives of the Jennings family. Carol, now in her mid-sixties, has Alzheimer's and 'essentially she's almost lost language now', says her son. The family began to notice subtle changes in her behaviour and ability to cope around 2004. John had just gone up to college and his parents were moving from their home in Lancaster to Warwickshire, where his father Stuart, a Methodist minister, had a new posting as university chaplain. The family had moved many times with Stuart's job and was well used to the process. But this time was different. 'Dad thought it was a bit odd because mum would generally be very organised, have all the boxes labelled,' says John. 'But when the removal company turned up on the day, things weren't really packed and it was all a bit chaotic. That was the first sort of sign.'

Despite the family's history of dementia, no one was really prepared for Carol's decline – not even Carol herself. Though she had given blood as part of the research process, she had never asked for the results. She thought she might 'collapse in a heap' if she knew for sure she was carrying the gene, she once told an interviewer. 'Mum's attitude was always: you could get hit by a bus tomorrow. That became almost her mantra,' says her son. 'That kind of denial carried on into her developing symptoms. And because she had this really resolute feeling about not wanting to know, people sort of tiptoed round mentioning it.'

But this was a selective taboo, says John. Alzheimer's was talked about often, and he even remembers writing poems about it at primary school. But the talk was always abstract and impersonal. 'It totally was abstract for me, because I hadn't seen the transition from a normal, functional adult to the granddad that I knew. I hadn't seen how the disease progressed, so I wasn't able to imagine it was possible for that to happen. It wasn't until mum started developing symptoms, with me as an adult, that it really dawned on me.'

There was just one occasion when John and his mother ventured into the difficult personal territory. Every summer when he and his sister were children their parents would rent a cottage in Northumberland for a family holiday, and John continued to join them there even after he'd grown up and left home. 'You know they say people are either larks or owls?' he commented. 'Mum and I were definitely owls, and so dad would go to bed and we would stay up late and have conversations. We had done so ever since I was a teenager, and we did a similar sort of thing this time, with a glass of wine. We spoke about the fact that we both thought we had the gene, and what it might be like [to develop dementia]. Mum said she thought it might be a bit like a television going to static, and there are probably worse things that can happen, as long as people are looking after you. That was unusual,' he continued, 'because we didn't ever speak about the possibility of *us* having it before then. And we didn't talk about it more after that – that was a one-off.'

Like his mother, John has resisted learning the results of his own blood test, but feels in his bones that he is carrying the mutant gene. He is actively involved in an international research programme called DIAN (Dominantly Inherited Alzheimer Network). Every two years he goes in for a battery of tests that include a lumbar puncture, blood tests and MRI and CT scans, and he, like his mother and several other family members, intends leaving his brain to science when he dies. The DIAN researchers have also been able to track the disease as it has developed in Carol Jennings, since Hardy, Rossor and their team began conducting MRI scans of her brain every year from the moment they discovered the mutant APP gene in 1991.

'One thing they have discovered from the DIAN project,' commented John, 'is that the deposition of amyloid starts long, long before the onset of symptoms – decades before – which means that if I do have the gene, that's already happening in me.'

In the search for drugs to treat Alzheimer's, Big Pharma has focused a good deal on amyloid – at clearing it from the brain, or preventing it from forming plaques. John Jennings follows the search for a drug with keen interest, but also a sense of fatalism. 'I've come to sort of just accept the narrative I've built for myself, that things are getting worse,' he laughs apologetically. 'That it's just going to speed up and then that will be it. Oblivion. Even if it's true that we can stop the disease by stopping amyloid, it's too late for me now,' he continued. 'Mum always said research was moving ahead so fast, but in terms of human lifespans it doesn't move fast enough!'

In fact, one clinical trial after another targeting amyloid has failed, either because the drug has proved to be unsafe, or because it just didn't work. This led Hardy, in a review of Alzheimer's research to mark the 100th anniversary of the

German psychiatrist's lecture, to conclude: 'If Auguste D. were alive today, her sad prognosis would be much the same as in 1906.'

But rather than defeat, this comment signals a sense of mighty frustration that in spite of all that has been learnt about Alzheimer's disease since Auguste Deter first aroused the curiosity of her psychiatrist, the Alzheimer nut is proving so very hard to crack. Some big questions remain to be answered. How, for example, do neurons die as the brain shrinks? Is it the tau tangles or the amyloid plaques – or something else altogether – that kills them? This is still today the subject of hot debate between the rival 'Tauists' and the 'BAPTists' (adherents to the 'beta-amyloid protein theory') – apt expressions because it has indeed been likened to a holy war between the competing theories as to which is the most important target for drugs.

Another big question is what the amyloid precursor protein – the product of the APP gene – does when it's not mutated. 'You know, we don't know!' says Hardy with an apologetic laugh. 'We know it's something to do with synaptic contacts, but honestly it's an embarrassment. We've known the gene for 30 years now and we really don't know what the protein does with any degree of certainty.'

And a third big question: why is it that some people who, in life, showed no signs of dementia are found at post-mortem to have brains clogged with amyloid? A famous example is Sister Mary, a participant in a long-running research programme among Roman Catholic sisters in the US that began in 1986. 'The Nun Study', as it was called, was set up by David Snowdon at the University of Minnesota, and was designed to observe the onset and progression of Alzheimer's disease over time. Snowdon and his fellow researchers recruited 678 sisters of a single religious order in the belief that studying a relatively homogenous group with a shared lifestyle would limit the variables that might confound their findings. Before her death in 1993 at the age of 101, reports Snowdon, Sister Mary achieved a high score in her cognitive tests, yet the post-mortem examination of her brain showed an abundance of plaques and tangles.

This last has been one of the big challenges to the amyloid hypothesis, constantly brought up by its critics, and there are no ready answers. 'It's humbling,' says Rossor. 'It reminds us that it's not just all very simple. Rarely in biology or medicine do you get nice one-to-one relationships.' A lot of these things are probabilistic, he explains, like smoking and lung cancer, high cholesterol and heart disease, obesity and diabetes – we all know people who beat the odds, but they don't disprove the links. 'Because there's so much else at play, you don't expect a one-to-one relationship, but on balance there is one.' While the story of Sister Mary and many others like her does give pause for thought, there is no escaping the fact that 'if you have a mutation in the APP gene, you end up with the full panoply of changes that we choose to call Alzheimer's disease', says Rossor.

These important questions are a reminder that much of our understanding of Alzheimer's comes from the study of the familial disease, where the relationship between genetic risk factor and outcome is pretty direct. Yet familial Alzheimer's accounts for only 2–3 per cent of all cases. The overwhelming majority of patients suffer from 'sporadic' disease, which means it's much less predictable – a matter of chance, time and lifestyle, as much as genetics. So what is known about this, the most widespread form of Alzheimer's?

CHAPTER EIGHTEEN

Alzheimer's disease – a challenge to amyloid

Allen Roses, maverick neurologist at Duke University, North Carolina, dropped dead of a heart attack at Kennedy International Airport on his way to a medical conference in Greece in 2016. He was 73 years old, and had been the bane of the amyloid cascade hypothesis for the past 23 years.

Roses was born in Paterson, New Jersey, in 1943, the son of Jewish immigrants from Poland who had escaped the Holocaust. His father Morris, who ran a stationery store, died when Allen was 13, and the teenager took on odd jobs to help support his family, including taking bets for bookies with dubious connections to the Mafia. Frequently described as 'pugnacious', he is also remembered by colleagues, family and friends as loving a good argument, and having a sense of humour and fun. 'Allen was a larger than life character, whose principal delight was taking the position opposite of conventional wisdom,' commented John Hardy in an obituary for the medical journal *The Lancet*. 'Usually this meant he was wrong, but his gadfly approach sometimes paid off with big findings.'

Principal among these big findings was the discovery in 1993 of a gene at the heart of sporadic (or late-onset) Alzheimer's disease called APOE e4. The story goes that in 1991, Warren Strittmatter, an Alzheimer's researcher in Roses's lab at Duke, was looking in the spinal fluid that bathes the brain for proteins that can bind with beta-amyloid, the stuff of the gunky plaques. When he found one that formed extremely strong bonds with the plaque protein in his test tube, his interest was piqued and he took his findings to his lab boss. 'It was like a lightning bolt,' Roses recalled many years later. The gene for the protein APOE was located on chromosome 19, which his team had already identified as important in genetic linkage studies among families with sporadic Alzheimer's.

But Roses found it impossible to get postdocs in his lab to investigate the gene further as they were all busy with what they reckoned was the brightest path to enlightenment – research into amyloid. So he did a deal with his wife Ann Saunders, a geneticist with exactly the skills needed to look into APOE but who was on maternity leave after the birth of their first child: Roses would look after the baby if she did the lab work. Within three weeks Saunders had data that strongly supported a role for APOE e4 in sporadic Alzheimer's disease. This is now known to be the single biggest genetic risk factor for the disease in well over half of all cases, leading Hardy in recent years to recognise Strittmatter and Roses's discovery as 'clearly perhaps the most important risk allele (gene variant) in the human genome'.

We humans have three common variants of APOE, which is a protein involved in packaging cholesterol and other fats and distributing them to wherever they are needed in the body. Numbered e2, e3 and e4, the variants have different levels of efficiency and differing effects on the brain, where cell membranes are composed largely of fats. This effect is dependent on what combination of the two alleles, or copies, of the gene you inherit (one each from mum and dad). APOE e2 seems to be protective against dementia. But if you inherit one copy of APOE e4 in combination with one of the others, your risk of developing Alzheimer's is two to three times greater than normal, and if both copies of the gene are APOE e4, the risk is roughly twelvefold. People with two copies of e4 are also likely to develop the disease earlier than those with other variants, often in their late sixties. Of the three variants, e3 is by far the most prevalent in the population, being roughly five times more common than the dodgy e4, and 10 times more common than the protective variant, e2.

These are the kind of statistics for risk you'll come across most commonly. But drill down a little deeper and the picture gets more complicated, since sex and race seem to have an influence. In 2014, for example, a group of scientists at Stanford University in California, intrigued by an old paper from the mid-1990s that suggested a sex difference in the effect of APOE e4 that had been largely overlooked by subsequent researchers, decided to investigate for themselves. Brain imaging had already provided evidence that this was probably a real phenomenon when it revealed that among healthy women with no signs of dementia, there were distinctive differences in the connectivity (or 'wiring') of the brains between those who carry the APOE e4 gene and those who do not. The scan images showed very little difference between the two groups of men.

For the new study, the scientists, led by Stanford neurologist Michael Greicius, looked at a very large pool of data gathered by dementia centres across the US that had followed clients across a seven-year period to 2013, testing their cognitive function annually. Of roughly 8,000 participants aged 60 and upwards picked for the Stanford analysis, about one-third had been assessed as having 'mild cognitive impairment' (MCI) at the start of the study, and two-thirds were without symptoms. The researchers found that, of the healthy group, women who carried a copy of APOE e4 were almost twice as likely as women without the dodgy gene to have developed symptoms by the end of the seven years, while in the men it seemed to make little difference. Among all participants who started the study with a diagnosis of MCI, however, having a copy of the dodgy gene increased the risk of going on to develop Alzheimer's equally for both women and men. Interestingly, the group at least risk of developing any symptoms of mental decline were the healthy women without an APOE e4 gene – which means there was a greater proportion of men who developed dementia that was totally unrelated to their APOE status.

As far as race is concerned, the evidence suggests that all combinations of the APOE gene that increase risk in Caucasians have a weaker effect in African Americans and Hispanics, and that the effects are strongest among people of Japanese origin. For a Japanese person with two copies of the e4 variant, for example, the risk of developing Alzheimer's is more than twice that of a Caucasian with the same two dodgy alleles. Interestingly, several studies in Nigeria have found that people carrying an APOE e4 gene – even those with two copies of this variant – are at no greater risk than anyone else of developing Alzheimer's disease. Why? We'll return to these conundrums a little later.

From the moment his lab discovered the risky gene, Roses challenged the amyloid hypothesis of Alzheimer's disease: plaques, he claimed, were a red herring, a consequence of the death of neurons, not a cause. With his students and in public lectures, he liked to draw the analogy of a graveyard with its tombstones marking the position of dead bodies. Nobody assumes the tombstones are what killed the person buried beneath them, he'd point out. Likewise, plaques 'are absolute markers required for the diagnosis of Alzheimer's disease. But that doesn't mean that that's what causes Alzheimer's disease.'

Roses didn't, at the beginning, have a clear alternative hypothesis for the cause of Alzheimer's. His challenge to amyloid may well have influenced the reception to his discovery of APOE e4. 'When Allen Roses said that gene was a main culprit, nobody took any notice, because it didn't fit the picture,' commented Ram Rao, who studies Alzheimer's disease at the Buck Institute in California, when I visited him at the bright, modern hilltop research centre. 'It was not sexy (I'm sorry to use that word)! You write anything about APOE e4 at the time, you won't get grants. And you go out at conferences and you didn't talk about A-beta and tau, you were ignored. Nobody wanted to listen …'

Roses was thwarted at every turn in his effort to get funding for further investigation, and in the end he sank his own money – to the tune of around half a million dollars, according to some reports – into his research. In 1997, a disillusioned Roses left academia for industry to carry on his work into APOE at what is now GlaxoSmithKline. He returned to Duke University 10 years later to run a drug-discovery programme looking for potential new medicines.

Much still remains to be learnt about how exactly APOE – whose basic task, as we've seen, is to shuttle fats around the body – works, and how it affects susceptibility to Alzheimer's. But more details of its relationship with amyloid have been uncovered since Strittmatter's first observations in the test tube. It seems that the APOE proteins have a profound influence on the fate of amyloid in between neurons – whether it transforms into the type that can aggregate as sticky fibres into plaques, or gets cleared away as part of routine housekeeping in the brain – with the three APOE variants having differential effects. In summary, e4 seems to promote the build-up of beta-amyloid, while e2 encourages its clearance and e3 doesn't have much impact. APOE also seems to influence the picture in other ways, including having a role in creating and maintaining synapses – the gaps between neurons across which signals pass – and in inflammation of the brain.

One theory is that the e4 variant leads to death of brain cells by interfering with the way mitochondria – the cells' batteries – metabolise glucose, thereby starving neurons of energy. This is the 'mitochondrial dysfunction' hypothesis. The story here is that, when stressed, brain cells produce APOE protein, and the rogue variant alone has the habit of breaking up into fragments that get into the nucleus and damage the batteries. At the Buck Institute in California, Ram Rao has done research that suggests APOE e4's ability to enter the nucleus – where our DNA is stored and where APOE has no right to be under normal circumstances – is also the key to the formation of the tau tangles.

'One of the things that happens to tau in Alzheimer's disease is it undergoes a change called phosphorylation,' says Rao. This means it gets tagged with a little molecule of phosphate that changes its behaviour 'And the more densely it's phosphorylated – we call it hyperphosphorylation – the greater the effect.' Under these conditions, the tau that's holding together the 'train tracks' of the microtubules, the brain's transport network, weakens and the tracks collapse.

Usually, the phosphate molecules tagging tau are removed regularly by an enzyme, says Rao. 'But we are now showing that APOE e4 actually directly affects this guy, the enzyme that removes the phosphate, so it's no longer functioning, it's aberrant. And so you have accumulation of the tau. And because of the accumulation you have the microtubules falling apart.' Rao's research also suggests that after the break-in to the nucleus, APOE e4 attaches itself to the DNA and takes control of the on/off switches of a large number of the genes besides tau that are involved in Alzheimer's. These include genes that control levels of the hormone oestrogen and may be what accounts for women's extra vulnerability to dementia, he says.

However, despite huge amounts of effort and much tinkering around with test tubes, transgenic mouse models and human materials, nothing is clear, and no one yet knows which mechanism, or mechanisms, might be most important. Nor can they even agree whether APOE e4 contributes to the development of Alzheimer's through *failure* to do what APOE should, or by actively doing something it should not, as Rao's and others' research suggests. (The very idea of APOE e4 breaking into the powerhouse of the cell is rubbished by some Alzheimer's scientists. 'That sounds like a crazy idea,' scoffed one researcher with whom I raised the issue. 'I think they smoke a bit too much dope down there [in California], to be honest!')

If this debate sounds like quibbling, it's not; settling the question of whether APOE e4 plays an active or passive role matters desperately in the search for effective drugs. That's why it excites such passions.

In 2009, Roses's team back at Duke discovered another gene, TOMM40, which sits right next to APOE on chromosome 19, and which they claimed has a role to play in the Alzheimer's story, sometimes in cooperation with APOE and sometimes alone. This gene is the recipe for a protein that's responsible for making channels through the outer membrane of mitochondria for the passage of all kinds of molecules vital to the functioning of the cells' batteries. Some versions of TOMM40 seem to disrupt this process and inhibit mitochondria in their job of producing energy, thus lending some support to the mitochondrial dysfunction hypothesis of Alzheimer's.

Roses's lab has also found that certain versions of the TOMM40 gene can influence the age of onset of Alzheimer's disease through interaction with APOE e3 – yes, precisely the APOE variant normally considered to be a negligible risk for Alzheimer's. This finding, however, is controversial. Some other labs, including that of Alison Goate – who, you'll recall from the previous chapter, was involved in finding the APP gene in the Jennings family – were not able to replicate Roses's results and are unconvinced about TOMM40's significance.

However, in late 2017, scientists working on a joint project between the University of Southern California (USC) and the University of Manchester, UK, found evidence that some variants of TOMM40 alone may be risk factors for loss of memory in older people. The press release put out by USC on publication of their findings – *Is the Alzheimer's gene the ring leader or the sidekick?* – was possibly intended to be provocative, given the mixed feelings that still prevail about TOMM40.

To look for genes associated with memory, the scientists had mined a rich seam of data gathered over the two decades up to 2012 by two long-term research projects – the US Health and Retirement Study and the English Longitudinal Study of Ageing. Participants in both studies had been given verbal tests to assess their immediate and delayed recall every two years as part of routine data collection, and their genetic make-up was also on record. Picking through the wealth of data for candidates who met their criteria of age, sex and race (they chose people with primarily European heritage to minimise race bias in their overall findings), the California and Manchester teams ended up with a total of nearly 14,500 people aged at least 50 years when their records began. The scientists ran the participants' memory test results against a battery of 1.2 million gene variants from across the human genome.

Though APOE e4 did show up as significant, both on its own and in conjunction with TOMM40, the gene that stood out most strongly from this massive soup of data for its effect on memory was TOMM40. 'Our findings indicate that TOMM40 plays a larger role, specifically, in decline of verbal learning after age 60,' the scientists wrote in their report. 'Further, our analysis showed that there are unique effects of TOMM40 beyond APOE e4 effects on both level of delayed recall prior to age 60 and decline in immediate recall after 60.'

Carol Prescott of USC, one of the 70 scientists involved in the research programme, commented, 'The results from this study provide more evidence that the causes of memory decline are even more complicated than we thought before. And they raise the question of how many findings in other studies have been attributed to APOE e4 that may be due to TOMM40 or a combination of TOMM40 and APOE e4.'

Whatever the answer to that, APOE and TOMM40 are just two of a growing number of genes that have been found to influence our susceptibility to dementia – some 25–30 for Alzheimer's disease alone – as technological advances have made the search ever easier. Today, hundreds of individual genomes can be sequenced and compared to give results in a matter of months. Hardy's lab in London is one of many around the world dedicated to finding 'the full panoply' of genetic contributors to dementia. And while he and his fellow geneticists identify the offending genes, others investigate what the genes do. They use models ranging from individual brain cells to fruit flies, and even very recently brains-in-a-dish – tiny bundles of brain tissue created from reprogrammed skin cells.

Most of the genes identified thus far as risk factors for sporadic Alzheimer's are 'housekeeping' genes, says

Hardy. 'They're directly involved in microglial activation – microglia are the little cells in the brain which basically clean up all the debris, all the damaged neurons.' Some mutations inhibit the function of the genes, 'so people with those variants just don't clean up the damage as well as they should – that's the essence of it.'

But just how much of a risk are these susceptibility genes to us as individuals? How concerned should we be if we have our genomes sequenced – as more and more of us seem to be doing – and find we're carrying one or a few of them? Each has only a tiny effect, says Hardy. Even the dominant one, APOE e4, has a small effect when compared with the three genes involved in familial Alzheimer's. To put it in context, 'if you have a single APOE e4 allele you have a threefold increase in your risk. Whereas if you have an amyloid mutation, you're going to get the disease,' he said. 'But in fact overall more people get Alzheimer's disease because of their APOE status than anything else, because 15 per cent of the population have an e4 allele. These APP mutations are rare as hen's teeth.'

A bunch of scientists at various universities in the US working with colleagues in Sweden were interested in quantifying the heritable risk of sporadic Alzheimer's more precisely. With psychology professor Margaret Gatz of the University of Southern California at the helm, they set up a study using the classic method for teasing apart genetic effects from other variables in our lives – twin studies. Such studies compare people who share either 100 per cent of their DNA in the case of identical twins, or 50 per cent in the case of non-identical twins, to see how likely or unlikely both members of a pair are to develop a particular condition if one of them does.

For their Alzheimer's study the scientists turned to the Swedish Twin Registry, a huge database collected over many years and including detailed information about environmental influences that may or may not be shared between the twins in a pair. This gave them a sample of nearly 12,000 twin pairs over the age of 65 years to work with, and in 2006 they published their findings – namely, that genes accounted for between 58 per cent and 79 per cent of the risk for Alzheimer's disease among their participants, with the effect being equally strong for men as for women. The researchers also found that genes seemed to influence the age of onset of dementia, with both twins who suffered from Alzheimer's in a pair tending to develop the disease around the same age.

The Swedish twin study, however, is not the last word on heritability of Alzheimer's disease; some other studies have found the genetic effect to be less strong. But whatever the true figure, the message is clear: that except in rare circumstances, our genes are *not* our destiny. What, then, are the other risk factors for this disease?

CHAPTER NINETEEN

It's the environment, stupid

'My own feeling is that, well, maybe somewhere in the range of half of Alzheimer's has an environmental component,' says Caleb Finch, a Professor of Neuroscience at the University of Southern California (USC). 'Tuck' Finch, as he's known, is the grand old man of the neurobiology of ageing. And, indeed, a pioneer. When he came into the field in 1965 there wasn't really a field to speak of, he told me as we sat together over lunch on the campus of the university where he's spent most of his working life.

As an undergraduate at Yale in 1959, Finch was considering going into developmental biology as a newish field with an exciting future. But one of his mentors at Yale, Carl Woese – a microbiologist who was to revolutionise our understanding of the biological tree of life with his discovery of a group of microbes called archaea that added a third 'branch' – suggested the other end of life might offer bigger challenges. 'He said if you really want to start a new field, why don't you think of ageing?' recalled Finch.

'At grad school I wound up doing my PhD thesis on ageing, and I decided that the brain was the major player in that ... I made a career commitment at that point, 1965, when I laid out this framework to myself – I decided that was going to be my life's work.' And nothing since has weakened his resolve, not even the dismissive remarks of the eminent virologist Peyton Rous, whom, you'll remember, we met earlier pouring cold water on Leonard Hayflick's seminal discovery of the finite life of dividing cells. When Finch gave a public talk on his PhD research into the ageing brain, he remembers Rous suggesting he was wasting his time – after all, everybody knew ageing was just about vascular disease and cancer.

Now in his late seventies, Finch is a tall, spare-framed, slightly stooping figure, balding, with a bushy grey beard and a lively, curious mind. 'He looks like God,' commented one of his former grad students for a profile in *Science* magazine. 'He looks like he came down from the Appalachians last week.' This last is appropriate, for Finch is a fiddle-player in his spare time, and a one-time member of the Iron Mountain String Band, which he founded in 1963 with a friend, Eric Davidson, a developmental biologist. Having learnt the trumpet at elementary school, he taught himself to play the traditional Appalachian fiddle at the age of 22.

Before the two scientists met, at grad school in New York City, Davidson had been collecting traditional music down in North Carolina and Southwest Virginia for the Smithsonian Library of Congress. Finch subsequently joined him, spending a week or so twice a year out in the field, collecting tunes on their heavy old reel-to-reel recording equipment. 'You'd go to a town, go to the barber's shop, go to the hardware store and say, "Who's an old-time fiddler or banjo player round here?" You'd chase down people and visit their homes, set things up and record,' recalled Finch. 'That's the music that we modelled the band on – the traditional southern Appalachian string band before bluegrass.' Few of the people they met on their road trips knew what they did for their day jobs, laughed Finch as he reminisced under the shadow of a huge, gnarled old Moreton Bay fig tree at the campus café, where he tucked into a plate of hamburger and chips.

As a scientist, Finch takes an unusually wide-angle approach to his subject. 'What I'm doing,' he explained, 'is sketching the horizon that very few of my peers or colleagues in biomedical gerontology are paying any attention to – the environmental side of ageing, which is, in my view, for humans, far more important than genetic variation.' The environmental influence has been largely ignored, he believes, 'because it's hard to study. You have to have a completely different set of premises and way of thinking about it that doesn't come out of the classical reductionism of biochemistry and molecular biology. That's a brilliant operating strategy for those sets of problems, but this is where the frontier is, in my view.'

Finch's own training was very different from the norm. As an undergraduate at Yale he was given a job as a lab assistant in the newly established department of biophysics, where, he says, 'there was a brilliant group of physicists coming into biology who were asking questions nobody else was asking. So that was my start. I had great early mentors who schooled me to be uninhibited about asking annoying questions.' He says that their attitude was '"Don't worry if nobody's done it before; it doesn't mean that it wasn't worth doing … leave those papers … Let's just take a broader look at what's going on in living systems that makes them different from physical systems." That was my training.'

Finch is interested today in how diseases of ageing have changed in the last 200 years as lifespans have increased. He is asking in particular whether our modern environment is exacerbating age-related diseases that may have been relatively rare in pre-industrial times. Teaming up with a multidisciplinary group of anthropologists and biomedical scientists, he has been studying the Tsimane people of the Bolivian Amazon, who until very recently lived as they always have done, by hunting, gathering, fishing and farming, beyond the reach of modern medicine or other amenities. 'They have a very high inflammatory load. They all have parasites. They have tuberculosis, and get sick frequently in their demanding day's work,' says Finch. 'You would predict – since inflammation drives many diseases – that they would have a high incidence of heart attacks, but they don't.'

Over the years of this long-running research programme – known as the Tsimane Health and Life History Project – cardiologists on the team have undertaken CT scans and electrocardiogram readings on hundreds of Tsimane participants. They have found that calcification of their arteries as they age occurs at a very much slower rate than it does among people living in the modern world – such that an 80-year-old Tsimane man typically has the vascular age of an American in his mid-fifties. We've imaged their brains too,' said Finch, 'and their rate of grey matter loss during ageing is at least 50 per cent slower than among people in North America and Europe.'

A mass of data has also been collected over the years on the Amazonian people's cognitive function, and this is beginning to reveal some very interesting findings about the interaction of genes and environment. Among the Tsimane, APOE e4 – considered the highest single risk factor for Alzheimer's disease in the industrialised world, as we have seen – appears to protect the brains of those who have a high burden of parasitic infections. What's more, this benefit seems to kick in very early: Tsimane children who carry a copy of the APOE e4 gene are generally brighter than those who don't. This mirrors findings from other studies among kids in poor communities in Mexico City and Brazil, who are especially vulnerable to infection, and where those carrying the e4 gene variant seem to have better cognition.

But the balance between the bugs and the genes and the impact on brains is complex. Uncontrolled parasitic infection damages the brain in its own right, so Tsimane of all ages who do not carry the protective e4 variant of APOE are vulnerable. And in those rare individuals who manage somehow to avoid parasitic infections, and yet who carry the APOE e4 gene, this variant behaves as it tends to in the modern world – it increases their risk of mental decline. These findings offer a plausible explanation, says Finch, for why a gene variant deemed to be such bad news has stuck around in the human population and not been weeded out by natural selection: it has a positive role to play in people who live as we all did for thousands of years in pre-industrial times, in intimate contact with a rich soup of invasive organisms. They may also be the clue to the race differences in the effect of APOE e4 mentioned earlier.

Harking back to our caveman origins, neuroscientist Ram Rao, who studies APOE at the Buck Institute, endorses Finch's interpretation of why the harmful e4 variant has persisted in the human population. 'This is a great story,' he says enthusiastically. 'APOE causes inflammation. And what happens is the caveman was always looking out for food. He didn't have shoes, socks and slippers. The caveman walked on bare ground, he climbed trees; he had to walk for miles just to get a good hunt which he could bring back to the family. In the course of all this he developed infections, he had bleeding [cuts and abrasions], and then he had to survive for a long period of time without food, if he didn't get the hunt. All that required for him to be active – APOE kept him active. So APOE is a good guy … it prevented infection from injuries spreading in the body.

'Now the same caveman, with all the medical interventions, lived beyond 45–50, started wearing shoes and trousers and shirts and eating all kinds of foods that are not good for the body, and APOE e4 is now in the system and doesn't know what to do, gets confused. And to use the Jekyll and Hyde story, it starts showing up its Hyde form. The same APOE e4 now triggers inflammation. Inflammation was good in the beginning, but now, as you become older, this inflammation goes awry.'

But what's going on between the genes and the bugs at the cellular level, deep within the Amazon people's brains? Poring over the rich accumulation of data brought back from their jungle field trips, the researchers have developed the theory that APOE e4 confers protection to the Tsimane via two possible mechanisms: by neutralising and clearing away parasites, or by changing the metabolism of cholesterol in the brain to alleviate the impact of the parasitic infection. But theories are just the beginning; there's a massive task ahead to confirm or refute them, and to pin down the mechanisms in fine detail.

The project among pre-industrial people offers tantalising evidence of the role of the environment in how our brains age. But what of the modern world with its many layers of protection against bugs and germs? What are the environmental threats here? The list of potential factors is enormous, says Finch, but his major focus at the moment is air pollution. We're talking here about ultra-fine particles measuring no more than 2.5 microns (that's about 30 times smaller in diameter than a human hair). Known as PM2.5s (shorthand for particulate matter 2.5), these are produced by burning fossil fuels and are pumped out mainly by power stations and the exhaust pipes of motor vehicles. They consist of a catalogue of nasties, such as sulphate, nitrate, hydrocarbons and heavy metals like lead, nickel and mercury. Evidence has been building for some time – and becoming more and more compelling – that air pollution can damage brains.

In the early 2000s, for instance, researchers in Mexico City – identified by the World Health Organization as one of the most smog-choked places on Earth – began monitoring the effects of the foul air on dogs, which share their environment with their owners, to get an idea of what air pollution might be doing to the city's human inhabitants. According to team leader Dr Lilian Calderón-Garcidueñas, people living in the study area reported signs of deranged behaviour in local dogs, such as altered sleep patterns and barking. Some owners told the researchers there were times when their dogs seemed not to recognise them. Examining the brains, post-mortem, of the dogs they were actively monitoring, the investigators found accumulations of beta-amyloid, plaques and other pathology reminiscent of Alzheimer's disease in humans, including death of neurons. In the concluding paragraph of their report, published in 2003 in the journal *Toxicologic Pathology*, they write: 'These canine findings are of sufficient magnitude and clinical significance to warrant concern that similar pathology may be occurring at an accelerated pace in humans residing in large metropolitan areas or those exposed to significant amounts of PM (particulate matter) such as the result of wildfires, disasters, or war events. Neurodegenerative disorders such as Alzheimer's disease may relate to air pollutant exposures.'

Links have been made between mental decline in elderly people and exposure to fine particle pollutants in a number of more recent studies in the US and elsewhere. Keen to pin down this gathering body of circumstantial evidence – to demonstrate cause and effect – Finch and his colleagues have been working on a project that combines human epidemiological studies with laboratory-based experiments with mice and cell cultures. They set out to answer three broad questions: do older people living in places with high levels of PM2.5 in the air have an increased risk of dementia; are those with the APOE e4 gene more sensitive to the effects of these pollutants; and could the findings in humans be reproduced under controlled conditions in the lab in mice genetically engineered to carry variants of APOE genes? If so, they reckoned, 'it could shed light on possible mechanisms underlying what is happening in human brains.'

For the human part of the project, Finch and his USC colleague, epidemiologist Jiu-Chiuan Chen, collaborated with the Women's Health Initiative Memory Study (WHIMS) run by researchers at Wake Forest University medical school, North Carolina. From their database they selected a sample of 3,647 women, who had been followed by the study since the late 1990s and who, at recruitment, were aged 65 to 79 and showed no signs of mental impairment. The women came from locations across the US. For all the participants, WHIMS had detailed information about their physical characteristics, clinical histories, lifestyles and behaviour, as well as their genetic profiles, revealing, importantly, their APOE status. The USC team used this rich resource alongside air-quality data gathered from the US Environmental Protection Agency to build a mathematical model that allowed them to estimate the everyday outdoor levels of PM2.5s in the different locations across a 10-year period to 2010, and thus the likely exposures of the women to harmful pollutants.

As all the pieces of the puzzle fell into place, the researchers saw that the women who lived in places where air-pollution levels regularly exceeded national safety standards had a much faster rate of mental deterioration than those living in less polluted environments, and that they were nearly twice as likely to develop dementia, including Alzheimer's. What's more, the risk for those carrying an APOE e4 gene was two to three times higher than for those with the other variants of the gene. 'If our results are applicable to the general population,' commented Finch, 'fine particulate pollution in the ambient air may be responsible for about one out of every five cases of dementia.'

Back in the lab, he and his group have exposed mice engineered to carry copies of the human APOE gene variants to carefully controlled doses of the superfine pollutants, PM2.5s, collected from motor traffic trundling the busy highway past the USC campus. They collaborated with Constantinos Sioutas from the USC engineering school who designed a sophisticated contraption of tubes and filters that capture the exhaust fumes for storage in liquid suspension. They can then be re-aerosolised for exposing the mice in the lab. 'That's a huge step above sticking cages of mice near a freeway, right?' commented Finch.

Over a period of 15 weeks, half the mice were exposed to the exhaust fumes for an average of five hours a day, three days a week. The other half – the control mice – were allowed to breathe clean air. All were then sacrificed and their brains examined and compared. The investigators found a great deal of inflammation caused by the microglia, the scavenger cells of the brain's immune system, activated to fight the invading particles. They found, too, high levels of an inflammatory molecule released by the microglia called TNF-alpha (tumour necrosis factor), which is typically elevated in the brains of people with Alzheimer's disease and linked to memory loss. Much as had Lilian Calderón-Garcidueñas in her dog study in Mexico City, Finch's group also saw excessive accumulation of beta-amyloid in the brains of the exposed mice. To analyse what was going on at a molecular level more definitively, they cultured some cells from the brain's immune system separately in lab dishes, and exposed them to exhaust fumes also.

'We now understand that particles generated by fossil fuels enter the body directly through the nose to the brain, as well as through the lungs, in our circulation,' said Finch in a news release from USC. 'And these have the net effect of causing inflammatory responses that increase the risk of Alzheimer's, and actually accelerate the process itself.' With his lab's studies of genetically engineered mice, he continued, 'we are able to definitively show that exposure to air pollution increases the level of brain amyloid – and to a greater extent in mice carrying the human Alzheimer risk factor, APOE e4.'

Our brains are protected from microbes and other harmful stuff circulating in the bloodstream by what's known as the blood–brain barrier, a tightly packed, semipermeable layer of endothelial cells lining the walls of the brain's blood vessels. But we know, says Finch, that this barrier is more porous than usual in people with APOE e4, thus enhancing this pathway into the brain for the superfine particles breathed in. The particles that enter the brain directly through the nose travel along the olfactory nerve, which is responsible for our sense of smell, and which connects, among other areas, to the hippocampus, where memories are laid down.

The olfactory nerve which runs from the nasal cavity is the only natural gap in the blood–brain barrier, since the ability to pick up scents instantly – and thus clues from our environment – is more or less vital to the survival of a species. Dogs have a much more acute sense of smell than we do, and in Calderón-Garcidueñas's Mexico study she found extensive damage to the olfactory system, from the nose to the brain. Interestingly, the inability to detect certain odours has recently been identified as an early sign of Alzheimer's disease, though the main mechanism seems to be clumping of beta-amyloid that kills olfactory nerve cells.

At USC, Finch is also very interested in how smoking affects the picture presented by air pollution, and here it's worth a slight detour to sketch in the background. Tobacco use has long been associated with an extra risk of cardiovascular disease and cancer. But until about 2010 the case for a link with dementia was hotly contested: some studies showed that smoking did increase the risk, while others found no effect, or that the risk was actually decreased. Then in 2010, Janine Cataldo and colleagues at the University of California, San Francisco, published a paper describing how they had systematically analysed the design, methodologies and findings of 43 original, international studies carried out between 1984 and 2009 to try to settle the matter. Significantly, they also looked at the funding and affiliation of the scientists involved to try to eliminate potential conflict of interest in the results – something that, surprisingly, seems to have been overlooked by the journals that published their reports. What was not surprising, however, to anyone who had followed the lengths to which the tobacco industry went to discredit the evidence of a link between smoking and cancer, was that Cataldo and her colleagues found Big Tobacco's fingerprints all over the evidence concerning Alzheimer's.

Discovering which bits of research and which scientists the industry had supported was a major sleuthing job. It involved trawling through a massive pile of previously secret internal papers held in the Legacy Tobacco Documents Library that was forced to open its archives when it faced litigation brought by irate customers for personal damages and deaths from smoking. The researchers found that 11 of the 43 studies included in their meta-analysis had been conducted by scientists with links to Big Tobacco, of which only three had disclosed their affiliation. None of the 11 studies found an increased risk of Alzheimer's among smokers – indeed, eight of them found a *decreased* risk of dementia, while the others found no significant effect. After adjusting for bias in the industry-supported studies and for some other factors such as study design, however, Cataldo and her group concluded that 'smoking is not protective against AD'. Indeed, the available data indicate that 'it is a significant and substantial risk factor for Alzheimer's disease'.

So what do the data mean for an individual who smokes? This obviously depends on the number of cigarettes the person smokes in a day, how long he or she has been a smoker, their genetic background and a host of other variables. However, a fact sheet from the World Health Organization dated 2014 cites a number of studies from around the world that put the added risk at between 59 per cent and 79 per cent. WHO estimates, furthermore, that around 14 per cent of Alzheimer's cases worldwide are 'potentially attributable' to use of tobacco.

Tobacco, says Finch, has many different mechanisms by which it increases or accelerates the risk of cardiovascular disease and cancer. But what about the brain? He is exploring the crossover between smoking and air pollution, looking at the extent to which they share mechanisms of action, and at whether or not these two types of bad air work synergistically to compound the risk of Alzheimer's. From the evidence available so far he says, 'I conclude that there's another level of injury that happens in the combination that hasn't been widely appreciated – and we *do not* have a mechanism to understand that.

'To take another perspective ... In enlightened countries smoking is being squeezed down to 10–15 per cent of adults. But every one of those adults, most of them live with other people, and so the total level of exposure within the home to second-hand smoke might be approaching 40–50 per cent of households. So even non-smokers in the household get a double hit, if they are unfortunate – as a third of the people in the world are – to live in a high-pollution zone.'

In the effort to identify risks for Alzheimer's in our modern world, air pollution of all sorts – be it from smoking, smog or anything else that involves breathing in nano-sized particles – is a new frontier, says Finch. His own research has raised important questions for which he is keen to find answers. For example, does air pollution *initiate* or simply accelerate the development of Alzheimer's disease? And are the results of his extensive studies in the females of the species applicable to males also?

Dementia, 'the Big D', is today one of the most feared aspects of growing old – largely because it still seems so indiscriminate, so relentless and so personally devastating. But what are the prospects for treatment looking like a century and a bit since Auguste Deter was taken to see a psychiatrist? 'I think that there's a tremendous opportunity to reduce the global burden of dementia dramatically,' says neurologist Dale Bredesen, a founding director of the Buck Institute back in 1999, and a practising physician.

Why is Bredesen so full of optimism? In the next chapters I'll look at the prospects today for the treatment of Alzheimer's disease. I'll look, too, at where the investigation into the other aspects of ageing – from shortening telomeres, senescent cells and deranged immunity to the Jekyll and Hyde face of some of our genes, and the destructive forces of free radicals – is leading in practical terms. Can anything yet be done to slow down or significantly ameliorate the process of ageing itself in us human beings?

CHAPTER TWENTY

Treat the person, not the disease

'We're beginning to understand the underlying biology, and we have to believe that through understanding the underlying biology we will eventually get to treatment,' said John Hardy, a little ruefully. 'But as someone said to me, "You scientists have been promising us jam tomorrow for an awfully long time." And that's true.'

We were sitting together in Hardy's office at UCL just days after the giant pharmaceutical company Eli Lilly had announced the failure of their anti-amyloid drug, solanezumab, in clinical trials. This was just the latest in a catalogue of crushing disappointments for patients and scientists alike: more than two decades of seeking a cure or preventative for Alzheimer's by targeting amyloid have yielded next to nothing. Drugs have been developed that harness the immune system to clear the sticky gunk from brains or to stop it accumulating in the first place, but if they haven't been abandoned in early trials because of safety concerns, most have shown no significant benefit to patients.

So did the failure of solanezumab suggest the drug companies have been barking up the wrong tree with their anti-amyloid drugs? 'Some people will say that. But I think they haven't been barking quite high enough!' said Hardy. Given the fact that the build-up of the faulty protein begins years – even decades – before any symptoms of dementia develop, 'I suspect it's failed because it was given too late. That's what I suspect.' Nor does Tuck Finch see the repeated failure of anti-amyloid drugs in clinical trials as a decisive challenge to the central importance of the plaques and tangles in Alzheimer's, but rather as a reflection of the enormous technical difficulty of 'moving from a biochemical understanding of the disease to a drug that is useful'.

Another amyloid-targeting drug, crenezumab, is still being trialled in rural Colombia, among members of a huge extended family of poor farmers in which the early-onset Alzheimer's gene presenilin 1 has been spreading for over 300 years. With more than 5,000 members, this is the largest known group of people with familial Alzheimer's in the world. Carriers of the fateful gene typically begin to show signs of mental decline in their mid-forties and to be lost in a wordless fog of dementia by the age of 50.

The trial, which gives stricken families hope of an escape from a disease known locally as *la bobera*, 'the foolishness', is administering crenezumab (or a placebo) to people who have inherited the faulty gene but who are typically years from developing symptoms. It is due to run for five years and end around 2021. But the failure thus far of so many anti-amyloid drugs has led to Alzheimer's being dubbed 'the graveyard of hope', and to the announcement in 2017 by the pharmaceutical giant Pfizer that it was abandoning dementia research – and thousands of jobs in neuroscience – to put its money into more promising ventures.

However, while the focus of Big Pharma has been on developing drugs to combat Alzheimer's, others have been looking elsewhere for solutions. 'This is a complex disease; you can't use a pill – one small pill – to address a chronic disease of this nature in the brain … There's no way on earth that we can improve the disease with a pill,' Ram Rao of the Buck Institute tells me with the passion of conviction. 'That's where NIH is putting all the money. If you talk about anything else beyond a pill approach they will say, "No, that doesn't work. That's not conventional Western medicine."'

Rao grew up in India, where he obtained his PhD in neuroscience before leaving for the US with a fellowship to work as a postdoc at the Mayo Clinic School of Medicine.

There he continued his studies into neurotransmitters, the brain's signalling chemicals, before moving to the Buck in California, one of its earliest recruits, appointed by the Institute's first president, neurologist Dale Bredesen. At one point, Rao's wife began to suffer a number of health problems from which she could find no relief. Eventually losing faith in 'Western' medicine, she decided to find herself an Ayurvedic doctor – a practitioner of the ancient Indian system of medicine. Ayurveda takes a holistic approach to healing, which means focusing on the whole person – body, mind, spirit and emotions – rather than on any individual organ or body part that is diseased.

Before her first appointment, Rao accompanied his wife to Sacramento to meet the doctor at one of the few Ayurvedic hospitals that existed in the US in the early 2000s. After lengthy discussions with the couple, the doctor suggested that Rao himself study Ayurveda. 'I said no, no … I am a Western guy, I study biochemistry; there is no question of me doing Ayurveda.' But when a programme started in San Francisco not far from the Buck, he was persuaded to attend the training course, and it changed his life.

Rao remains a pure scientist, focused in the lab on the minutiae of Alzheimer's, on what's going on in the cells. But today he is also an Ayurvedic practitioner who, like Tuck Finch, believes in looking at the bigger picture, at what is happening to the whole person when they develop dementia. 'For example, here I'm looking at APOE e4, one small, single molecule,' he explains. 'But now what I'm doing is stepping back and saying, okay, it's doing so many things; how can I address all these pathways simultaneously?'

His primary focus is on reducing chronic inflammation in the brain, and he's looking especially at the role the gut is playing in this condition. Leaky guts – which allow normally friendly bacteria to escape into the bloodstream where they are seen as aliens and keep the immune system on its toes – are, he believes, as much of a problem to the brain as they are to the rest of the body. All kinds of things can weaken the gut walls, including poor diet, erratic eating habits and stress. All three together are especially bad for us, says Rao, because they play havoc with our digestive systems, and have a knock-on effect on the biochemistry of our brains – which means that what, when and how we eat is vital to brain health. 'Most of the stuff I talk about is not pulling things from a hat,' he says. 'I'm just putting two and two together and I'm backing it up with evidence-based research. I know the Ayurvedic concepts; I know the Western concepts; I know they mix with each other, but now, to present it to others, I need to have the evidence-based research, and that's what I'm doing.'

Standing at the interface between the two systems, Rao notes that modern science is validating some of the core practices of traditional Indian medicine. Besides the focus on eating habits to preserve a healthy gut, these include the stress-busting strategies of yoga and meditation. Ayurveda also emphasises the delivery of medicines directly through the nose by inhalation (known as 'Nasya'). Traditional practitioners have been exploiting this direct route to the brain without a scientific foundation, he says. 'Now Big Pharma is looking at sending individual molecules into the brain through the nose.'

A 2012 paper from a group at Washington University, Seattle, led by Suzanne Craft describes their programme to test this route for the delivery of insulin to treat people with early Alzheimer's disease, or mild cognitive impairment – considered a step on the way to dementia. Alzheimer's is sometimes referred to as 'diabetes of the brain' because insulin deficiency or resistance – and the consequent inability to metabolise glucose effectively to fuel the cells – are common features. But giving insulin the usual way, by injection, to non-diabetic elderly people risks serious side effects from plummeting blood-sugar levels. Insulin delivered directly to the starved neurons through the nose, the researchers found, significantly improved the memory, thinking and learning abilities of all the participants in their study, while the brains of the people who received the placebo continued to deteriorate. Encouraged by the results of one of the first trials in humans, the researchers suggest this might be an effective strategy for stabilising or slowing the advance of Alzheimer's by boosting the energy output of brain cells before too many of them have died.

Dale Bredesen is taking a lead in proposing a change of paradigm to reduce the global burden of dementia. A clinical neurologist by training, as well as a research scientist with a distinguished record of guddling around at the most fundamental level of the brain to work out how the machinery goes wrong, Bredesen has, since 2012, been treating his patients in a novel way. He starts from the premise that while the defining hallmarks of Alzheimer's – the plaques and tangles – have a decisive role, they are by no means the cause of the disease; they are instead protective responses to some insult to the brain such as being starved of vital nutrients, invaded by microbes, exposed to toxic substances, or some combination of these.

Seen for decades as the bad guy, beta-amyloid has in recent years been shown to have antimicrobial properties. What happens with Alzheimer's, says Bredesen, is that the brain is literally trying to protect itself from things that would damage it. 'A-beta is a little bit like napalm,' he explains. 'If you've got people breaching your borders, then you're going to try to kill them. But in so doing you're putting down something that is toxic and reduces your arable soil, so you're now living in a smaller country.' The analogy is deliberately shocking, for what he believes has happened to the brains of people with Alzheimer's is that 'they've downsized their overall network as they were attempting to kill the invaders'. In other words, reduced regions of their brains to rubble in trying to protect it.

Simply clearing away the responding agents, the amyloid or tau – the napalm of the analogy – is not going to solve the problem. The important thing is to identify and deal with the original insult to the brain that prompted the response. Thus the first thing Bredesen does with a new patient is to run a battery of tests; he takes blood samples and arranges for a series of scans that will give him a picture of what's going on generally in important systems of the patient's body. The tests monitor their digestive systems, microbiomes (the colonies of bacteria found in their guts and in their noses with which they live in symbiosis, more or less harmoniously), DNA (noting particularly their APOE status), exposure to environmental toxins (smoking and air pollution, for example), markers of inflammation, and their hormone balance. In all, 'there are over a hundred different determinants, and it's helpful to know, where does this person stand? Especially with respect to APOE status – because it does matter in terms of how you deal with people,' Bredesen said during a presentation of his programme to the Silicon Valley Health Institute.

For most people who go to their doctors worried they may be losing their minds, the chances of getting such a thorough assessment of their health is next to nil. 'There's this feeling in the neurological community that Alzheimer's is divorced from everything going on in the rest of your body – which I think is just crazy,' said Bredesen. The battery of tests he runs generates such a wealth of information that he uses a computer to analyse exactly what is going on in a patient and to help in developing a remedial programme tailor-made to the individual and designed to correct all the bits of the system that are out of kilter. He illustrates his multiple-front approach with the analogy of a roof full of holes that all need patching if they're to keep the rain out. 'If you are going to help yourself ... If you take a drug for Alzheimer's, it patches one hole; it patches the hole beautifully, but it's only one hole. So we want to patch all the holes – and the good news is we *can* patch all the holes,' he assured his Silicon Valley audience.

At its most basic, Bredesen sees Alzheimer's as caused by an imbalance between the two opposing processes that give the brain its 'plasticity' – that is, its ability to modify its connections and rewire itself constantly as we go through life from infancy to maturity and beyond. Each neuron in the adult brain has around 10,000–15,000 synapses, or connections to other neurons; these are constantly being built, maintained and repaired, as well as pruned and remodelled by complementary 'synaptoblastic' and 'synaptoclastic' processes.

The therapeutic programme he devises for his patients seeks to make adjustments to the myriad things that play into these two processes. It includes dietary instructions – what to eat (lots of fruit and veg and non-farmed fish) and what to avoid as far as possible (e.g. gluten, refined white flour, sugar, processed foods and red meat); a fistful of vitamins and other dietary supplements; regular exercise; a recommended 7–8 hours' sleep per night; periods of fasting between meals (specifically, at least three hours between supper and bedtime, and 12 hours before breakfast); stress-busting activities like yoga and meditation; and brain stimulation with mental exercises and games.

The regimen demands long-term commitment to some serious lifestyle changes and pill-popping, and is not easy to stick to, admits Bredesen. But he's had some remarkable successes with patients who do manage to follow most of his prescriptions. He cites the case of a 67-year-old woman – in fact his first client, 'Patient Zero' – whose mother had developed Alzheimer's in her mid-sixties and died of the disease, and who had been led to expect the same fate when she went to her GP fretting about memory loss. She had become disorientated when driving, she told him, and often could not remember where to turn off the freeway even on well-known routes. She muddled up the names of her pets, and forgot where the light switches were in her own home. And she was fast losing her ability to cope with her work – she had a demanding job which involved data gathering and analysis for the government, lots of foreign travel, and report writing. 'She simply couldn't crunch the data any more,' said Bredesen, and she had been considering suicide before she came to his clinic. But just three months into his programme, she had called to say she felt better than she had in many years, and was back at work full time and coping fine. Another patient who had felt her mind slipping away before she started her programme told him some time later that she had allowed herself to start talking to her grandchildren again about the future.

Bredesen's casebook contains many such examples, but, as the neurologist himself acknowledges, the evidence that his approach is effective is mostly anecdotal thus far, says fellow neuroscientist Tuck Finch. There is a pressing need, he says, for large-scale, carefully constructed trials that provide such objective data as 'before and after' images of amyloid and tau pathology, and much else.

Finch is, however, much more open to ideas from left field than most scientists, and Bredesen faces stiff resistance from many mainstream neuroscience colleagues. 'I served on the National Aging Council; we published over 220 papers; we had millions of dollars of grants,' he told me. However, 'once we started to say something different, and actually be able to make a difference for human beings …

Oh my gosh, I couldn't get a grant if my life depended on it! We are seen as mavericks. And yet we have the only results to date that have made human beings better.'

The reason for the hostile reception to his ideas, he believes, is because 'we're really completely changing the paradigm, and we're saying, look, you guys are *wrong*. This is not about making a toxic peptide, A-beta. This is all about a response to different insults.' His own change of direction, he explains, came as a result of frustration with the snail's pace of progress in combatting Alzheimer's – the long decades at the bench that had teased out the genetic basis of dementia and revealed the hallmarks of Alzheimer's, but left a yawning gulf of understanding between the two points. In tackling that yawning gulf, Bredesen, like Ram Rao, began to see wisdom in the ancient traditions of Ayurveda and Chinese medicine, with their holistic view of disease offering a counterbalance to the reductionism of molecular biology.

'I would say that "aha" moment came when we were studying the amyloid precursor protein signalling [in my lab] and we could see that it was literally functioning as a molecular switch – you could push it in one direction, you could push it in the other direction ... You know, making synapses, reorganising synapses. That's when we realised, okay we need to look at *all* the things that push this switch in each direction.' Bredesen realised, too, that this kind of approach was central to Ayurvedic medicine. 'When I was at medical school I thought this stuff was just stuff that was done thousands of years ago because they didn't know any better – it really didn't have much of an effect,' he told me. 'But I realise now, no, it had a tremendous effect! They were on the right path.' The task ahead, he says, is 'to convince our neurologists that these neurodegenerative diseases are *systemic* problems with neurological read-outs – they're not isolated brain problems.'

Two small clinical trials currently underway are aimed at doing just that, and at challenging the conviction that there's little or nothing we can do at present for people who are losing their minds.

We have been focusing on Alzheimer's disease as one of the more frightening manifestations of old age. Research is going on into how to relieve many of the other manifestations too. But in the final chapter we want to step back and look at the bigger picture, for the ultimate aim of geroscience is to find a treatment for ageing itself.

CHAPTER TWENTY ONE

Ageing research – from the lab into our lives

You'll have heard the one about dark chocolate being good for you. And the one about red wine being the answer to the 'French Paradox' – the question of why the French have such low incidence of heart disease (apparently), despite their rich diet. The 'mystery' ingredient behind both these urban myths is resveratrol, a compound made naturally by many plants – notably red grapes, blueberries, mulberries, cranberries and peanuts – to fight off microbial invaders and fungi.

The claims made for resveratrol in chocolate and red wine are urban myths because you'd have to consume inherently harmful amounts of the stuff to achieve a meaningful dose. Nevertheless, resveratrol is an ingredient in traditional Chinese and Japanese medicine, and researchers have been interested in it since the early 1990s, when two plant scientists at Cornell University in the US first suggested it might be responsible for France's healthy hearts, thereby stimulating a flurry of overblown claims in the media about the benefits of a bit of dietary self-indulgence.

Despite the hype, scientific interest was properly piqued; labs around the world got stuck into research, and resveratrol was soon found to extend the lifespan of model organisms from yeast, fruit flies and worms to mice and fish. The compound has been studied extensively for its potential anticancer and antioxidant properties in humans, and for its effects on metabolism, blood flow to the brain, and of course its effects on the heart. One researcher was so excited by his results with yeast that he started taking resveratrol supplements himself – even gave it to his family – and in 2004 he set up a small biotech company to develop resveratrol-based drugs to prevent a host of age-related conditions and hopefully extend human life. The company, Sirtris, was bought by GlaxoSmithKline in 2008, but wound up in 2013 because of lack of progress, doubts about the compound's modes of action, and safety concerns – in clinical trials, some people developed nausea, diarrhoea and kidney problems.

But interest in resveratrol didn't die along with the company. The take-home message from Sirtris's failure was that a lot more work was needed on understanding how the compound really works and on refining the ingredients with the desired effects. In 2017, three UK-based gerontologists – Lorna Harries of Exeter University, with Richard Faragher and Lizzy Ostler of Brighton University – announced the result of their work with compounds precisely fashioned from the 'blunt tool' of resveratrol: the three had managed to bring senescent cells back to younthful function. The scientists were looking at what are known as 'RNA splicing factors' – scraps of protein within cells that act like minuscule scissors, editing the ribbon of instructions sent out by an activated gene to the cellular machinery to make the protein that will carry out its work. Splicing factors become increasingly sloppy in their editing tasks as we age, so that the instructions reaching the protein-making machinery are ever less precise and the function of genes – and therefore the activity of cells – is compromised. This is thought to contribute to the frailty of elderly people and to other age-related diseases.

The waning efficiency, and even loss altogether, of some RNA splicing factors as the genes that produce them stop working properly is a particular feature of senescent cells.

Could this deficit be corrected, the researchers wondered, and to what effect? Resveratrol is known to affect many different cellular mechanisms, including RNA splicing, so the team designed compounds based on the natural product to target this mechanism preferentially, and applied them to senescent cells in culture. The results were remarkable. 'I couldn't believe it,' said Eva Latorre, who did most of the work at the lab bench in Exeter. 'These old cells were looking like young cells. It was like magic.' Latorre repeated her experiments a number of times to make sure of what she was seeing, but it was clear. Within hours, the flabby old cells had perked up, truncated telomeres had been repaired and the cells had started to grow again.

'This is a first step in trying to make people live normal lifespans, but with health for their entire life,' said Harries. 'Our data suggests that using chemicals to switch back on [the RNA splicing factor genes] that are switched off as we age might provide a means to restore function to old cells.'

But there is an immensely long and rocky road between exciting results in the lab and something in the medicine cupboard we humans can take to keep us healthy or make us better. An analysis of the drug development and approval process published in the journal of the Royal Pharmaceutical Society, for example, winds up with the sobering conclusion that 'for every 25,000 compounds that start in the laboratory, 25 are tested in humans, 5 make it to market and just one recoups what was invested'. The attrition rate is similar in the US, where the rules and regulations of the Food and Drug Administration (FDA) governing development and approval of medicines have got increasingly stringent across the years. By the mid-1990s, for instance, a typical new drug had to be tested on nearly 5,000 people in more than 60 clinical trials to gain approval, compared with around 30 clinical trials and 1,500 people in 1980. Only around one in 1,000 compounds that looks promising in the lab even gets the go-ahead in the first place to set off on the road to the clinic.

One way around the painfully slow and uncertain process of bringing a new drug to market is to look for drugs already in the medicine cupboard that might have wider applications than the diseases for which they were designed. There are some colourful examples of what's known as 'repurposing'. The drug zidovudine was developed in the mid-1960s as an anticancer agent. Then in 1985, when scientists were casting around for something to stop the fearsome spread of HIV, zidovudine was found to have antiretroviral properties. It became the first drug licensed to treat infection with the AIDS virus, hitting the market in record time – as 'AZT' in 1987 – under intense pressure from frantic AIDS activists.

Viagra, taken by millions of men worldwide as an aid to sexual performance, started life as a treatment for angina – chest pain associated with heart problems – but found new purpose swiftly when men taking part in clinical trials reported that it gave them strong and enduring erections (some participants apparently refused to hand back the drugs still in their possession when the trial ended).

And then there's thalidomide. The drug that caused serious birth defects in the late 1950s and '60s, when taken by pregnant women as a treatment for morning sickness, was originally developed as a sedative and sleeping pill. Despite its association with tragically blighted lives and acrimonious battles in the law courts, thalidomide is used today under various names to treat a complication of leprosy known as ENL (erythema nodosum leprosum), characterised by large, extremely painful boils and severe inflammation. Its effectiveness against this condition was found by pure chance. In a review article for *Nature* on drug repurposing, Ted Ashburn and Karl Thor recount how, in 1964, a doctor in France, Jacob Sheskin, scoured the medicine cupboard for something to give a leprosy patient in such severe pain from ENL that he hadn't slept for weeks. Coming upon thalidomide, he gave his patient a dose, and to his astonishment found that not only did the drug give the afflicted man a good night's sleep, but it seemed to clear his agonising ulcers. Sheskin subsequently confirmed the effectiveness of thalidomide in a double-blind trial with leprosy patients with ENL. Then in the mid-1990s, the drug was found to inhibit angiogenesis – the growth of blood vessels – and it is used today to treat certain cancers that depend on developing their own blood supply to survive and spread. Because of its extensive history of testing in clinical trials and its use with leprosy, securing FDA approval in 2012 for thalidomide to treat multiple myeloma, a cancer of white blood cells, is estimated to have cost US$40–80 million, compared with the US$2 billion and more it typically costs to bring a brand new drug to the market.

In just the last decade, repurposing of drugs that have already jumped many, if not all, of the hoops on the way to the medicine cupboard has really taken off. And this is the idea behind the first and so far only programme to test a drug for the treatment of ageing per se – something that will strike at the roots of the process itself rather than at any one of the individual diseases associated with ageing. The programme, known by its acronym TAME (Targeting Ageing with Metformin), was conceived during a brainstorming session by a bunch of geroscientists holed up in a medieval castle-turned-hotel in rural Spain in 2013 to discuss how to take the fruits of their research forward towards the clinic. Now the scientists are busy raising the funds for a placebo-controlled clinical trial of metformin – at present the world's most widely used antidiabetic drug – that will involve some 3,000 people aged 65–79 years and 14 centres of investigation across the US, and is due to last around six years. Half the participants will receive the active drug and half a placebo.

TAME, which has an impressive cast of gerontologists on its team, is led by Nir Barzilai, Director of the Institute for Aging Research at the Albert Einstein College of Medicine, New York – and, incidentally, the man behind the genetics study of supercentenarians among the Ashkenazi Jews in the Bronx, mentioned in Chapter 7. Born in Haifa in 1955 and raised in Israel, Barzilai is a small, stocky man with a mop of iron-grey hair, and eyes almost permanently crinkled in a smile behind heavy specs. He exudes enthusiasm, good humour and a can-do sense of purpose.

This last Barzilai attributes to his experiences as a medical officer in the Israeli Army. As a member of the special forces, he took part in the dramatic raid of July 1976 on Uganda's Entebbe Airport, launched to rescue 102 Israeli passengers held hostage on an Air France jet hijacked en route from Tel Aviv to Paris by two Palestinians and two members of a German left-wing group. Barzilai later served for a while as the Israeli Army's chief medical officer, carrying out his rounds by helicopter much of the time. Such experiences hold many lessons in living, he told *Science* magazine. 'The major thing is that you realise you can do a lot! If it doesn't frighten you, you can do a lot.'

Barzilai's interest in ageing as a phenomenon was stimulated very early in his life, he told me when I managed to corner this small human dynamo for a while at a gerontology conference in New York. 'When I was 13 I was walking with my grandfather ... I used to take a walk with him every Saturday morning, and he would tell me his stories of his youth. And I was thinking, "He can barely walk ... what's he *talking* about!"' he laughs. 'You know, they say that young people have imagination, and I guess they do in a sense. But when you see your grandparents, you don't see it as your fate, you know? You see it more as: "Oh, they must have looked something like that forever, and we're different, right?"'

Though age has always been an important reference point in treating patients, there wasn't much interest in ageing as a study topic when Barzilai qualified as a doctor. So when he took up a research fellowship at Yale in the late 1980s, he focused on metabolism, a system he knew changes dramatically as we age. One of the drugs he investigated for its effect on controlling blood sugar was metformin, little knowing then what an important role this drug would play in ageing research some 30 years later.

Metformin is derived from a lupin-like plant, *Galega officinalis*, commonly known as goat's rue or French lilac – a native of the Middle East and now naturalised across Europe and the western fringes of Asia, though classified as an invasive weed in the US. The plant has been used for centuries in folk medicine – often as a remedy for excessive urination, which is a tell-tale sign of diabetes – and its derivative metformin was first discovered in the 1920s to have sugar-lowering properties in rabbits. The first person to test metformin in humans was Jean Sterne, a French doctor and diabetes specialist, who published his impressive results with the drug in 1957. It was licensed for use in the UK the following year and successively in other countries. But though it took the US until 1994 finally to approve it, metformin is today the treatment of choice worldwide to control blood sugar in type 2 diabetes. The drug is a generic* costing a few pence or cents a dose, and tens of

* A generic drug is a pharmaceutical drug that is equivalent to a brand-name product in dosage, strength, route of administration, quality, performance and intended use, but is no longer under patent and does not carry the brand name.

thousands of tons of the pills are produced annually, mostly in pharmaceutical factories in India.

In recent years metformin began to demonstrate effectiveness way beyond diabetes. Researchers found that not only does it extend significantly the lifespan of model organisms from worms to rats and mice, but it also improves their health and vigour. And in 2014 a retrospective study in the UK suggested it has a similar effect in humans. The original goal of this study was to compare the effectiveness of metformin with another first-line drug for diabetes. The researchers used a huge data set of NHS information from clinical practice for the year 2000 to look at the survival rates of around 78,000 diabetics being treated with metformin, 12,000 diabetics on another first-line drug, and 90,500 carefully matched controls without diabetes. To their surprise they found that not only did the diabetics on metformin have a far greater chance of survival than those on the other drug, but they also did significantly better than the non-diabetic controls, suggesting that metformin had a general protective effect against the ravages of time.

Scientists still have a lot to learn about metformin's mechanisms of action, but they believe its principal role is basically to enhance the activity of an enzyme within cells that inhibits the process of burning glucose for energy – thereby mimicking the effects of calorie restriction, with its various benefits of reducing oxidative damage and inflammation. Acting on promising evidence from observations that it helps to prevent or control tumours, the cancer research community is already running a large number of clinical trials with the drug. And now, with the TAME study, it's being asked to strut its stuff on the even bigger stage of ageing.

Metformin was not, however, the only drug under consideration at that hotel in Spain for this groundbreaking programme. Steven Austad, who was on the team (and whom we met earlier taming lions and investigating long-lived creatures like Ming the mollusc), was in favour of using rapamycin, 'because the animal results have been so spectacular', he told *Science*. Though less impressive in many respects than rapamycin, metformin has a long, strong safety record, whereas rapamycin, the drug used by Lynne Cox and Judy Campisi to rejuvenate senescent cells in their labs, has some more or less serious side effects. 'Nir said, "We can't afford in this first trial to kill anybody," continued Austad. 'And I thought, "Strategically, he's right."' And strategy here was all important, since testing the effectiveness of the drug at delaying ageing is not the primary purpose of the TAME trial.

So what is the main purpose of the trial? As everything you've read so far testifies, gerontologists have made enormous progress in understanding what happens to our bodies as we grow older. But they face a big barrier in translating what they have learnt into anything clinically useful since ageing per se is not seen by most of us – certainly not by the drug agencies or health-insurance companies – as a disease, and not therefore as a legitimate target for intervention. With no defined market, there is no incentive for Big Pharma – the only player with any real clout – to get involved in drug development. What's needed to break the logjam, Barzilai and colleagues reckoned, is for a drug-regulating authority to specifically recognise ageing as a medical condition that can be modified – thereby delaying the onset of the miseries that plague the elderly and drain the healthcare budget. A paper in the journal *Health Affairs* suggested in 2013 that the US alone could save around US$7.1 trillion over 50 years (and buy individuals around 2.2 extra years of life into the bargain) by intervening in the process of ageing itself.

'The beauty of this kind of approach is that [it bypasses] this awful futility of treating ageing-related diseases,' comments David Gems, the UCL geneticist we originally met in Chapter 2. 'Firstly, once these diseases have formed, they're very difficult to treat; but also in a sense you're looking at a *syndrome* with ageing. My mum was a classic example of the relative futility of trying to treat individual diseases. Her health was very poor when she got older; she almost died from cardiovascular troubles, and they managed to adjust her drugs and bring her back from the edge; and she was fine for a bit but then she got breast cancer and dementia. So you basically treat one symptom and the other symptoms just take their place,' he continued. 'But if you're looking at the root causes of a whole spectrum of pathologies you push them all back – which is what you see in the animal models.'

In 2015, a core group of delegates from TAME that included Barzilai and Austad travelled to Silver Springs, Maryland, on the fringe of Washington DC, to put the case for their clinical trial to the FDA. Metformin, safe and familiar to everyone present, was to be their battering ram; their means to provide 'proof of principle' that ageing is a target worth aiming for. So important was their appointment with the FDA that the delegates – all academic scientists, without any representatives of Big Pharma – rehearsed their arguments intensively beforehand in a nearby hotel. The group agonised over how to describe their mission with metformin, given the association of ageing research with quacks and charlatans peddling dreams of immortality – and also the gut-resistance of most ordinary people to the pathologising of old age.

The answer, they decided, was to avoid mentioning ageing directly as the target of treatment, but to talk instead of 'co-morbidities' – in other words, to characterise ageing as a syndrome comprising a bunch of diseases that tend to occur in later life. As such, the delegates would tell the FDA, the TAME study will measure how quickly, if at all, individual participants newly develop one or more of the age-related diseases (heart problems, cancer, dementia) or die during the period of the trial. 'Even in our mind, in my mind, ageing is not a disease,' Barzilai told *Science* magazine soon after the meeting with the FDA. 'It's, you know, humanity! You're born, you die, you age in between ... I'm kind of saying, "I don't care what they want to call it, if I can delay it."'

The meeting with the FDA went surprisingly well. A goodly number of the agency's senior staff had gathered to hear the TAME presentation, and they were clearly impressed. The scientists left 90 minutes later with the agency's endorsement for the trial and its aims. But the final question of official recognition of ageing as an 'indication', or medical condition, for treatment – and thus one that healthcare services and insurers will be prepared to pay for – they agreed would await the results of the trial.

Once that final hurdle is jumped the floodgates will be open for Big Pharma to get involved – and the prospects for them will be extremely attractive. As the Deputy Director of the FDA, Robert Temple, commented after the presentation from TAME, 'If you really are doing something to alter ageing, the population of interest is everybody. It surely would be revolutionary if they can bring it off.'

A revolution, yes. But there will never be a single elixir of youth because we all respond differently to drugs, depending on our personal biology, our genetic background and our environmental exposure. What works brilliantly for some may work less well, or not at all, for others. This is the lesson from the convoluted highways and byways of ageing research explored in this book, if one takes a sober look beyond the eye-catching headlines.

Let me take you back for a minute to the Buck Institute in California, and to the office of Pankaj Kapahi on a summer's afternoon in 2016. We have been talking about dietary restriction and its dramatic effect on the lifespan of fruit flies in the lab, which can be double or even triple the norm. A big extension in lifespan is the dominant effect of the experiments with dietary restriction (DR), but what does the picture look like when you focus in closer? Kapahi takes me over to a couple of large posters on his office wall covered with blue and red dots. These are the results of experiments with DR using wild fruit flies collected in a local marketplace. They represent 200 different strains and huge genetic diversity. The 'red dots' have been allowed to feed at will on a nutrient-rich diet, while the 'blue dots' have been given frugal diets. Neither the reds nor the blues, representing individual flies, are bunched neatly but are scattered right across the graph, above and below the line representing the normal lifespan.

'If you were these guys,' says Kapahi, pointing at some blue dots, 'it's amazing! DR is doubling, sometimes tripling the lifespan. But if you're a guy here,' he points to a scatter of dots below the line, 'it's actually shortening the lifespan – the blue is going lower. All these are going lower! So this is what's going to happen with any intervention, right? You can't do this experiment on people, but we can do it on flies, and we can see the genetic variability is huge.'

Kapahi moves to another poster showing the effect of DR on the individual flies' energy levels, measured by disturbing test tubes of the little creatures and seeing how high and how long they jumped above a threshold. Once again the dots are widely scattered. Even among the longer-lived flies the effect is not uniform, underlining just how much a critter's personal biology affects the outcome of any intervention: in their experiments a longer life didn't automatically imply youthful vigour. Here the devil really is in the detail. The lesson is that the promise of anti-ageing therapy lies, as for all sophisticated healthcare procedures, with 'personalised medicine' – remedies tailored to us as individuals.

So when are we, ordinary citizens of the world, likely to see the benefit of all this anti-ageing research? It's the question I put to Kapahi's colleague at the Buck Institute, Gordon Lithgow, who coined the term 'geroscience' around 2006, and who has watched the story unfold from the front lines in labs on both sides of the Atlantic. 'Infectious diseases are the example I always use,' says Lithgow. 'I think we're at the moment Fleming was at when he discovered penicillin. So, he discovered it; he went to conferences; he talked about it; people said, "Hm, that's interesting." And there was about a 10-year gap, I think, between the discovery and the actual manufacture of penicillin.' The breakthrough came on the back of the realisation that infectious diseases had a common cause – microbes – and could be hit with a common strategy. This was a paradigm-shifting moment in biomedicine, says Lithgow, 'and it changed everything.'

And that's where we are again today: at another 'Fleming moment', with the discovery that the diseases of ageing have common roots, and that the process of ageing itself is not immutable. The task now, says Lithgow, is to wake up the policy-makers, people in government, in the health services and the health-insurance business, to these facts.

'It's not inevitable that people get Alzheimer's; it's not inevitable that people get cancer and heart disease,' he insists. 'If we invest in the science, we have a choice of going down a road that is different from the one we're on, which is to build long-term care facilities – you know, treat symptoms; manage horrible diseases ...' This is the equivalent, he says, of building sanatoriums, designing iron lungs. These things no longer exist because we have no need for them in an era of vaccines and antimicrobials.

'And we can go down a different road here so we're not building the iron lungs for ageing and the sanatoriums for ageing, we're actually preventing disease.'

The message is the same from Nir Barzilai, who also draws comparisons with infectious diseases. 'Today, the possibility of living healthier and longer [lives] is not science fiction—it's science,' he writes in a blog for TEDMED. 'The metformin clinical trial will serve as the tool or framework for the most important medical intervention in the modern era since antibiotics—a new category of drugs that add years of healthy life as we age.'

Barzilai concludes, 'With a concentrated, coordinated effort across the public, private, and philanthropic sectors, together we can move ageing research from the labs into our lives.'

Notes on sources

During my research for this book I have drawn on a wealth of materials from books, journals and multimedia presentations by or about scientists and their work for information, insights and ideas. Here I list just the key sources for each chapter. Those that have proved relevant to the discussion throughout the book, rather than simply to individual chapters, include:

Appleyard, Bryan, *How to Live Forever or Die Trying* (Simon & Schuster, London, 2007).

Austad, Stephen N., *Why We Age: What Science Is Discovering about the Body's Journey through Life* (John Wiley & Sons Ltd, New York, 1997).

Guarente, Lenny, *Ageless Quest* (Cold Spring Harbor Laboratory Press, New York, 2003).

Hall, Stephen S., *Merchants of Immortality* (Houghton Mifflin Company, New York, 2003).

Kirkwood, Tom, *Time of Our Lives* (Weidenfeld & Nicholson, London, 1999).

Magnusson, Sally, *Where the Memories Go* (Hodder & Stoughton Ltd, London, 2014).

Olshansky, S. Jay and Carnes, Bruce A., *The Quest for Immortality: Science at the Frontiers of Aging* (W. W. Norton & Co., New York, 2001).

Walker, Richard F., *Why We Age: Insight into the Cause of Growing Old* (Dove Medical Press, 2013).

Preface

The quote from Ezekiel Emanuel comes from 'Why I hope to die at 75', published in *The Atlantic*, October 2014. See: www.theatlantic.com/magazine/archive/2014/10/why-i-hope-to-die-at-75/379329.

The quote from the doctor in the NHS comes from comment section of *New Scientist* magazine, 16th December 2017.

Chapter 1: So what is ageing?

The opening paragraph of this chapter comes from the book *Why We Age: Insight into the Cause of Growing Old,* by Richard F. Walker (Dove Medical Press, 2013).

The hallmarks of ageing are taken from 'The Hallmarks of Aging', by Carlos López-Otín, Maria A. Blasco, Linda Partridge, Manuel Serrano and Guido Kroemer, published in *Cell* 153: 1194–1217 (2013). See: www.cell.com/fulltext/S0092-8674%2813%2900645-4.

For information on the history of ageing theories I drew on a number of sources, including:

The online Medicine Encyclopedia, 'Evolution of Aging – Antagonistic Pleiotropy Theory Of Aging ("Pay Later" Theory)', available at: medicine.jrank.org/pages/609/Evolution-Aging-Antagonistic-pleiotropy-theory-aging-pay-later-theory.html.

'Weismann's Programmed Death Theory', available at: www.programmed-aging.org/theories/weismann_programmed_death.html.

'Aristotle On Old Age', blog posted on 26th April 2008 by Professor Camillo Di Cicco, available at: www.science20.com/scientist/blog/aristotle_old_age-27964.

Principles of Bioenergetics, by Vladimir Skulachev, Alexander V. Bogachev and Felix O. Kasparinsky, published by Springer, 2013.

Chapter 2: Wear and tear?

A key source of information on free radicals was '"Heavy" fat – the secret to eternal youth?', by Jessica Hamzelou, in *New Scientist,* 16th May 2015.

The review paper by David Gems and Ryan Doonan quoted at the end is 'Antioxidant defense and aging in *C. elegans*'. In: *Cell Cycle* 8(11): 1681–1687 (2009), available at: www.ncbi.nlm.nih.gov/pubmed/19411855.

Chapter 3: Telomeres – measuring the lifetime of cells

A key resource for this chapter, and for anecdotes about Leonard Hayflick, was Stephen S. Hall's book, *Merchants of Immortality* (Houghton Mifflin Company, New York, 2003).

Hayflick's original letter of rejection for his paper from journal editor Peyton Rous can be seen at: www.michaelwest.org/aging-under-glass.htm.

Another key resource for this chapter was the archive of the Nobel Foundation, which awarded the Nobel Prize for Physiology or Medicine to Barbara McClintock in 1983, see: www.nobelprize.org/nobel_prizes/medicine/laureates/1983, and the same prize to Elizabeth Blackburn in 2009, see: www.nobelprize.org/nobel_prizes/medicine/laureates/2009.

Additional quotes from Blackburn come from *Discover* magazine, 6th December 2007. 'Scientist of the Year Notable: Elizabeth Blackburn', by Linda Marsa. See: www.discovermagazine.com/2007/dec/blackburn.

And also from her YouTube conversation with *iBiology*, see: www.youtube.com/watch?v=0zfpfD_ILF0.

Guarente, Lenny, *Ageless Quest* (Cold Spring Harbor Laboratory Press, New York, 2003).

Michael West tells his story in *The Translational Scientist* magazine, November 2016. See: www.thetranslationalscientist.com/issues/0816/lessons-ive-learned-with-michael-west.

Chapter 4: Cell senescence – down but not out

A key source for this chapter was Richard Faragher's presentation to the Royal Society in London on 7th June 2016. His talk – entitled 'Live longer, live well – seize the day!' – was part of the annual Successful Ageing Programme organised jointly by the European Dana Alliance for the Brain and The University of the Third Age. Available on YouTube at: www.youtube.com/watch?v=_prg77TVOQQ.

Information about Maximina Yun's work with salamanders is drawn mostly from her research paper, 'Conserved and novel functions of programmed cellular senescence during vertebrate development', by Hongorzul Davaapil, Jeremy P. Brockes and

Maximina H. Yun, published in *Development* 144(1): 106–114 (2017). See: www.ncbi.nlm.nih.gov/pubmed/27888193.

Chapter 6: Ming the Mollusc and other models

Key sources on *Drosophila* for this chapter were the archive of the Nobel Foundation, which awarded Thomas Hunt Morgan the Prize for Physiology or Medicine in 1933. See: www.nobelprize.org/nobel_prizes/medicine/laureates/1933/morgan-article.html.

J. V. Chamary's excellent feature article, 'Modern Biology Began in the New York "Fly Room"', published in *Forbes* magazine on 18th March 2016. See: www.forbes.com/sites/jvchamary/2016/03/18/the-fly-room/#2ff6909c306d.

And 'Fruit flies in the laboratory', published in YG Topics by *Your Genome*, available at: www.yourgenome.org/stories/fruit-flies-in-the-laboratory.

For information on *C. elegans* and the scientists who first studied it, I drew on Andrew Brown's excellent 'biography' of the little creature, *In the Beginning Was the Worm: Finding the Secrets of Life in a Tiny Hermaphrodite* (Simon & Schuster, 2003).

Chapter 7: It's in the genes

Key sources for this chapter were:

'In Methuselah's Mould', by Bill O'Neill, published in *PLOS Biology* 2(1): e12 (2004). See: doi.org/10.1371/journal.pbio.0020012.

'A method for the isolation of longevity mutants in the nematode *Caenorhabditis elegans* and initial results', by Michael J. Klass, published in *Mechanisms of Ageing and Development* 22(3–4): 279–286 (1983).

'A personal retrospective on the genetics of aging', by Thomas Johnson, published in *Biogerontology* 3: 7–12 (2002). See: link.springer.com/article/10.1023%2FA%3A1015270322517

Cynthia Kenyon's talk to TEDGlobal in July 2011, entitled 'Experiments that hint of longer lives', available at: www.ted.com/talks/cynthia_kenyon_experiments_that_hint_of_longer_lives.

'The first long-lived mutants: discovery of the insulin/IGF-1 pathway for ageing', by Cynthia Kenyon, published in the *Philosophical Transactions of the Royal Society B* 366: 9–16 (2011). See: rstb.royalsocietypublishing.org/content/366/1561/9.

'FOXO3A genotype is strongly associated with human longevity', by Bradley Willcox et al., published in the *Proceedings of the National Academy of Sciences* 105(37): 13987–13992 (2008). See: www.ncbi.nlm.nih.gov/pmc/articles/PMC2544566.

'Long live FOXO: unraveling the role of FOXO proteins in aging and longevity', by Rute Martins, Gordon J. Lithgow and Wolfgang Link, published in *Aging Cell* 15(2): 196–207 (2016). See: www.ncbi.nlm.nih.gov/pmc/articles/PMC4783344.

'Extension of life-span by loss of CHICO, a *Drosophila* insulin receptor substrate protein', by D. J. Clancy, D. Gems, L. G. Harshman, S. Oldham, H. Stocker, E. Hafen, S. J. Leevers and L. Partridge, published in *Science* 292: 104–106 (2001). See: pdfs.semanticscholar.org/1fe7/57bae7bf7c56e605e336896661728c86c5f7.pdf.

The information on the Boston group's supercentenarian study comes from the paper 'Health span approximates life span among many supercentenarians: compression of morbidity at the approximate limit of life span', by Stacy L. Andersen, Paola Sebastiani, Daniel A. Dworkis, Lori Feldman and Thomas T. Perls, published in a special issue on extreme longevity by *Journals of Gerontology: Series A, Biological Sciences and Medical Sciences* 67A(4): 395–405 (2012).

David Gems was interviewed by Kat Arney for *The Naked Scientist* podcast in August 2014. Available at: www.thenakedscientists.com/articles/interviews/prof-david-gems-healthy-ageing

Linda Partridge and David Gems present their thoughts on the 'hyperfunction theory' in their paper 'Genetics of longevity in model organisms: debates and paradigm shifts', published in *Annual Review of Physiology* 75: 621–644 (2013). See: www.ucl.ac.uk/~ucbtdag/Gems_2013.pdf.

Chapter 8: Eat less, live longer?

Key sources for this chapter were:

'Honoring Clive McCay and 75 years of calorie restriction research', by Roger B. McDonald and Jon J. Ramsey, published in

The Journal of Nutrition 140(7): 1205–1210 (2010). See: www.ncbi.nlm.nih.gov/pmc/articles/PMC2884327.

'Clive McCay: a man before his time', a commentary by William Hansel, published in *Endocrinology & Metabolic Syndrome* 5(3): 236 (2016). See: www.omicsonline.org/open-access/clive-mckay-a-man-before-his-time-2161-1017-1000236.php?aid=73637.

Obituaries to Roy Walford, among the most informative being: *The Independent* newspaper, UK: www.independent.co.uk/news/world/americas/diet-guru-who-tried-to-live-for-ever-bequeaths-spartan-regime-58978.html, the *Chicago Tribune*: articles.chicagotribune.com/2004-05-03/news/0405030078_1_dr-roy-walford-diet-span, and the *Los Angeles Times*: articles.latimes.com/2004/may/01/local/me-walford1.

'"Biospheric Medicine" as viewed from the two-year first closure of Biosphere 2', by Roy L. Walford, R. Bechtel, T. MacCullum, D. E. Paglia and L. J. Weber, published in *Aviation Space Environmental Medicine* 67: 609–617 (1996).

'Can extreme calorie counting make you live longer?', by Peter Bowes, published by *BBC News* on 11th February 2013. See: www.bbc.co.uk/news/magazine-21125016.

Linda Partridge's lecture titled 'Manipulating nutrient-sensing signalling to improve health during ageing', delivered at the Molecular Frontiers Symposium held in Gothenburg, Sweden, in August 2016. It can be viewed on YouTube: www.youtube.com/watch?v=cJhiNNyXBC0.

Chapter 9: The immune system – first responders

Key sources for this chapter were:

'Chronic inflammation (inflammaging) and its potential contribution to age-associated diseases', by Claudio Franceschi and Judith Campisi, published in *The Journals of Gerontology: Series A* 69(Suppl 1): S4–S9 (2014). See: doi.org/10.1093/gerona/glu057.

'Virus-induced NETs – critical component of host defense or pathogenic mediator?', by Craig Jenne and Paul Kubes, published in *PLOS Pathogens* 11(1): e1004546 (2015). See: journals.plos.org/plospathogens/article?id=10.1371/journal.ppat.1004546.

Chapter 10: The immune system – the specialists take over

A rich source of information and quotes for this chapter was Janko Nikolich-Zugich's lecture, 'Aging of the Immune System', delivered to the SENS Research Foundation in California on 15th January 2014. See: www.youtube.com/watch?v=VVbGGA7ze1c.

Chapter 11: The bugs fight back

Besides my interviews with Janko Nikolich-Zugich, much of the information in this chapter is drawn from the paper, 'Known unknowns: how might the persistent herpesvirome shape immunity and aging?', by Nikolich-Zugich, J., Goodrum, F., Knox, K. and Smithey, M. J., published in *Current Opinion in Immunology* 48: 23–30 (2017). See: www.ncbi.nlm.nih.gov/pubmed/28780492.

Chapter 12: HIV/AIDS – adding insult to injury

A useful source of information for this chapter was the review article, 'Is HIV a model of accelerated or accentuated aging?', by Sophia Pathai, Hendren Bajillan, Alan L. Landay and Kevin P. High, published in *Journals of Gerontology: Series A, Biological Sciences and Medical Sciences* 69(7): 833–842 (2014). See: www.ncbi.nlm.nih.gov/pubmed/24158766.

A useful source of historical information was *AIDS: Images of the Epidemic*, published by the World Health Organization in 1994. ISBN 9241561637.

Information on the SMART trial came from 'CD4+ count–guided interruption of antiretroviral treatment', published by The Strategies for Management of Antiretroviral Therapy (SMART) Study Group, in *New England Journal of Medicine* 355: 2283–2296 (2006). See: www.nejm.org/doi/full/10.1056/NEJMoa062360.

Information for the START trial came from 'Initiation of antiretroviral therapy in early asymptomatic HIV infection', published by The INSIGHT START Study Group, in *New England Journal of Medicine* 373: 795–807 (2015). See: https://www.nejm.org/doi/full/10.1056/NEJMoa1506816.

Chapter 13: Epigenetics and chronology – the two faces of time

The key sources of information on Steve Horvath's epigenetic clock were his original article, 'DNA methylation age of human tissues and cell types', published in *Genome Biology* 14(10): R115 (2013), available at: www.ncbi.nlm.nih.gov/pmc/articles/PMC4015143.

And the feature article, 'Biomarkers and ageing: The clock-watcher', by W. Wayt Gibbs, published by *Nature* on 8th April 2014. See: www.nature.com/news/biomarkers-and-ageing-the-clock-watcher-1.15014.

Background information on Wolf Reik and Oliver Stegle's epigenetic clock in mice was drawn from their feature article 'How epigenetics may help us slow down the ageing clock', published in *The Conversation* on 10th May 2017. See: www.theconversation.com/how-epigenetics-may-help-us-slow-down-the-ageing-clock-76878.

Chapter 14: Stem cells – back to fundamentals

The archive of the Nobel Foundation was a rich source of information about Shinya Yamanaka, who won the Prize for Physiology or Medicine in 2012. See: www.nobelprize.org/nobel_prizes/medicine/laureates/2012/yamanaka-bio.html.

Another rich resource for this chapter was a news feature, 'How iPS cells changed the world', by Megan Scudellari, published by *Nature* on 15th June 2016 See: www.nature.com/news/how-ips-cells-changed-the-world-1.20079.

Also, the Yamanaka group's own papers:

'Induction of pluripotent stem cells from mouse embryonic and adult fibroblast cultures by defined factors', by K. Takahashi and S. Yamanaka, published in *Cell* 126(4): 663–676 (2006).

'Induction of pluripotent stem cells from adult human fibroblasts by defined factors', by Kazutoshi Takahashi, Koji Tanabe, Mari Ohnuki, Megumi Narita, Tomoko Ichisaka, Kiichiro Tomoda and Shinya Yamanaka, published in *Cell* 131: 861–872 (2007). See: www.cell.com/cell/pdf/S0092-8674(07)01471-7.pdf.

For information about Juan Carlos Izpisua Belmonte and his work, I drew on the following sources: 'The creator of the pig-human chimera keeps proving other scientists wrong', by Usha Lee McFarling, published on 7th August 2017 by the online health-orientated news website *STAT*. See: www.statnews.com/2017/08/07/pig-human-chimera-izpisua-belmonte.

And the original paper from the Izpisua Belmonte group, 'In vivo amelioration of age-associated hallmarks by partial reprogramming', by Alejandro Ocampo et al., published in *Cell* 167: 1719–1733 (2016). See: www.ncbi.nlm.nih.gov/pmc/articles/PMC5679279.

The press release of their paper is available at: www.salk.edu/news-release/turning-back-time-salk-scientists-reverse-signs-aging.

For information on Hendrikje van Andel-Schipper I drew on two main sources: 'Blood of world's oldest woman hints at limits of life', by Andy Coghlan, published in *New Scientist* magazine on 23rd April 2014. See: www.newscientist.com/article/dn25458-blood-of-worlds-oldest-woman-hints-at-limits-of-life.

And 'In Old Blood', by Jef Akst, published by *The Scientist* on 1st August 2014. See: www.the-scientist.com/?articles.view/articleNo/40567/title/In-Old-Blood.

Chapter 15: Something in the blood?

For background information on parabiosis I drew heavily on the paper 'Heterochronic parabiosis: historical perspective and methodological considerations for studies of aging and longevity', by Michael J. Conboy, Irina M. Conboy and Thomas A. Rando, published in *Aging Cell* 12(3): 525–530 (2013). See: www.ncbi.nlm.nih.gov/pmc/articles/PMC4072458.

Another very useful source of information was 'Ageing research: Blood to blood', by Megan Scudellari, published in *Nature News*, 21st January 2015. See: www.nature.com/news/ageing-research-blood-to-blood-1.16762.

I drew also on Tony Wyss-Coray's excellent talk for TEDGlobal, 'How young blood might help reverse aging. Yes, really', delivered in June 2015. See: www.ted.com/talks/tony_wyss_coray_how_young_blood_might_help_reverse_aging_yes_really.

And the report 'Infusions of young blood tested in patients with dementia', by Alison Abbott, published in *Nature News* on 3rd November 2017. See: www.nature.com/news/infusions-of-young-blood-tested-in-patients-with-dementia-1.22930.

And the press release from the University of California, Berkeley, 'Young blood does not reverse aging in old mice', published in *Science Daily* on 22nd November 2016. See: www.sciencedaily.com/releases/2016/11/161122123102.htm.

Information for the parabiosis footnote comes from 'Parabiosis in mice: a detailed protocol', by Paniz Kamran, Konstantina-Ioanna Sereti, Peng Zhao, Shah R. Ali, Irving L. Weissman and Reza Ardehali, published by the *Journal of Visualized Experiments* 80: 50556 (2013). See: www.ncbi.nlm.nih.gov/pmc/articles/PMC3938334.

Chapter 16: The broken brain

A rich source of information for this chapter was 'A hundred years of Alzheimer's disease research', by John Hardy, published in the journal *Neuron* 52(1): 3–13 (2006). See: www.cell.com/neuron/fulltext/S0896-6273(06)00723-9.

Also the article 'Auguste D and Alzheimer's disease', by Konrad Maurer, Stephan Volk and Hector Gerbaldo, published in *The Lancet* 349: 1546–1549 (1997). See: www.thelancet.com/journals/lancet/article/PIIS0140-6736(96)10203-8/fulltext.

Another good source of historical information was the book *Concepts of Alzheimer Disease: Biological, Clinical, and Cultural Perspectives*, edited by Peter J. Whitehouse, Konrad Maurer and Jesse F. Ballenger, published by Johns Hopkins University Press, Baltimore, in May 2003.

Other key resources were:

The obituary to Martin Roth written by Claude M. Wischik and published in *The Independent* newspaper on 18th October 2006. See: www.independent.co.uk/news/obituaries/professor-sir-martin-roth-420652.html.

The obituary '"Father of AD Neuropathology" Sir Bernard Tomlinson dies at 96', published by *Alzforum* on 16th June 2017. See: www.alzforum.org/news/community-news/father-ad-neuropathology-sir-bernard-tomlinson-dies-96.

For information on Arvid Carlsson and tensions within the neuroscience community I drew on Carlsson's lecture to the Nobel committee in December 2000, held in the archive of the Nobel Foundation. See: www.nobelprize.org/nobel_prizes/medicine/laureates/2000/carlsson-lecture.html.

Chapter 17: Alzheimer's disease – the family that led the way

Besides my long interview with John Jennings, I drew for this chapter on a video interview with Carol Jennings posted on the website of University College London's dementia education/training programme, *FutureLearn*, in the category 'The Many Faces of Dementia'. See: www.futurelearn.com/courses/faces-of-dementia/0/steps/12922.

As with the previous chapter, I found a rich source of information in 'A hundred years of Alzheimer's disease research', by John Hardy, published in the journal *Neuron* 52(1): 3–13 (2006). See: www.cell.com/neuron/fulltext/S0896-6273(06)00723-9.

For the story about Sister Mary and the mystery of her amyloid-clogged brain I drew on 'Aging and Alzheimer's disease: lessons from the Nun Study', by David A. Snowdon, published in *The Gerontologist* 37(2): 150–156 (1997). See: academic.oup.com/gerontologist/article/37/2/150/616995.

Chapter 18: Alzheimer's disease – a challenge to amyloid

For information about the personal life of Allen Roses I drew on the excellent obituary 'Allen Roses, Who Upset Common Wisdom on Cause of Alzheimer's, Dies at 73', by Sam Roberts, published in *The New York Times* on 5th October 2016, see: www.nytimes.com/2016/10/06/science/allen-roses-who-upset-common-wisdom-on-cause-of-alzheimers-dies-at-73.html, and on the obituary by Alison Snyder published in *The Lancet* 388: 2232 (2016), see: www.thelancet.com/pdfs/journals/lancet/PIIS0140-6736(16)32081-5.pdf.

A key resource for this chapter was the news feature 'Alzheimer's disease: the forgetting gene', by Laura Spinney, published in *Nature* 510(7503): 26–28 (2014). See: www.nature.com/news/alzheimer-s-disease-the-forgetting-gene-1.15342.

Another important resource for this chapter was the paper 'ApoE and A-beta in Alzheimer's disease: accidental encounters or partners?', by Takahisa Kanekiyo, Huaxi Xu and Guojun Bu, published in *Neuron* 81(4): 740–754 (2014). See: www.ncbi.nlm.nih.gov/pmc/articles/PMC3983361.

For information on sexual differences in the effect of APOE, I drew on the paper 'Sex modifies the *APOE*-related risk of developing Alzheimer's disease', by Andre Altmann, Lu Tian, Victor W. Henderson and Michael D. Greicius, published in *Annals of Neurology* 75(4): 563–573 (2014). See: www.ncbi.nlm.nih.gov/pmc/articles/PMC4117990.

And also the news release from Stanford University Medical Center, 'Gene variant puts women at higher risk of Alzheimer's than it does men, study finds', published in *ScienceDaily*, 14th April 2014. See: www.sciencedaily.com/releases/2014/04/140414191451.htm.

Robert Finn's report 'Neuroscience Meeting To Feature Feisty Debate On Alzheimer's Etiology' from the annual meeting of the Society for Neuroscience, published by *The Scientist* on 16th October 1995, offered further colourful information about Allen Roses. See: www.the-scientist.com/?articles.view/articleNo/17606/title/Neuroscience-Meeting-To-Feature-Feisty-Debate-On-Alzheimer-s-Etiology.

For information about the relationship between APOE e4 and TOMM40 I drew on the news release from the University of Southern California, 'Is the Alzheimer's gene the ring leader or the sidekick?', published by *EurekAlert!* on 14th September 2017. See: www.eurekalert.org/pub_releases/2017-09/uosc-it091417.php.

The source of information on the twin study was the paper 'Role of genes and environments for explaining Alzheimer disease', by Margaret Gatz, Chandra A. Reynolds, Laura Fratiglioni, Boo Johansson, James A. Mortimer, Stig Berg, Amy Fiske and Nancy L. Pedersen, published in *Archives of General Psychiatry* 63: 168–174 (2006). See: jamanetwork.com/journals/jamapsychiatry/fullarticle/209307.

Chapter 19: It's the environment, stupid

Besides my own interview with Caleb Finch, I found useful personal information in the profile article 'Listening to the Song of Senescence', by Ingfei Chen, published in *Science* magazine on 5th February 2003. See: www.sciencemag.org/careers/2003/02/listening-song-senescence.

For information on the Tsimane studies, I drew on the journal papers 'Coronary atherosclerosis in indigenous South American Tsimane: a cross-sectional cohort study', by Hillard Kaplan et al., published in *The Lancet* 389: 1730–1739 (2017). See: www.ncbi.nlm.nih.gov/pubmed/28320601.

And 'Apolipoprotein E4 is associated with improved cognitive function in Amazonian forager-horticulturalists with a high parasite burden', by Benjamin C. Trumble, Jonathan Stieglitz, Aaron D. Blackwell, Hooman Allayee, Bret Beheim, Caleb E. Finch, Michael Gurven and Hillard Kaplan, published in *FASEB Journal* 31(4): 1508–1515 (2017). See: www.ncbi.nlm.nih.gov/pmc/articles/PMC5349792.

And the feature article 'An Ancient Cure for Alzheimer's?', by Pagan Kennedy, published in *The New York Times* on 14th July 2017. See: www.nytimes.com/2017/07/14/opinion/sunday/alzheimers-cure-south-america.html.

Other key resources for this chapter were 'The Terrifying Truth About Air Pollution and Dementia', by Aaron Reuben, published by *Mother Jones* on 24th June 2015. See: www.motherjones.com/environment/2015/06/air-pollution-dementia-alzheimers-brain.

'The Polluted Brain', by Emily Underwood, published by *Science* magazine on 26th January 2017. See: science.sciencemag.org/content/355/6323/342.

'This is the link between air pollution and dementia', by Caleb Finch and Jiu Chiuan Chen, published by the World Economic Forum in collaboration with *The Conversation* on 6th March 2017. See: www.weforum.org/agenda/2017/03/this-is-the-link-between-air-pollution-and-dementia.

'Cigarette smoking is a risk factor for Alzheimer's disease: An analysis controlling for tobacco industry affiliation', by Janine K. Cataldo, Judith J. Prochaska and Stanton A. Glantz, published in *Journal of Alzheimers Disease* 19(2): 465–480 (2010). See: www.ncbi.nlm.nih.gov/pmc/articles/PMC2906761.

For the Mexican dog study, I drew on the paper 'DNA damage in nasal and brain tissues of canines exposed to air pollutants is associated with evidence of chronic brain inflammation and neurodegeneration', by Lilian Calderón-Garcidueñas et al., published in *Toxicologic Pathology* 31: 524–538 (2003). See: journals.sagepub.com/doi/abs/10.1080/01926230390226645.

The quotation from Dale Bredesen in this chapter comes from an interview with him conducted by Chris Kresser for *Revolution Health Radio* on 14th July 2016. 'RHR: Prevention and Treatment of Alzheimer's from a Functional Perspective—With Dr. Dale Bredesen' is available at: chriskresser.com/prevention-and-treatment-of-alzheimers-from-a-functional-perspective-with-dr-dale-bredesen.

Chapter 20: Treat the person, not the disease

An important source for information on the familial Alzheimer's in Colombia was 'New Drug Trial Seeks to Stop Alzheimer's Before It Starts', by Pam Belluck, published by *The New York Times* on 15th May 2012. See: www.nytimes.com/2012/05/16/health/research/prevention-is-goal-of-alzheimers-drug-trial.html.

Besides my interview with him, information about Dale Bredesen's approach to treating Alzheimer's came from the following sources: 'Reversing Alzheimer's Disease', a presentation by Dale Bredesen delivered at the Silicon Valley Health Institute on 17th November 2016. See: www.youtube.com/watch?v=6D5aA_-3Ip8&t=156s.

And 'Reversal of cognitive decline: a novel therapeutic program', by Dale E. Bredesen, published in *Aging (Albany NY)* 6(9): 707–717 (2014). See: www.ncbi.nlm.nih.gov/pmc/articles/PMC4221920.

Chapter 21: Ageing research – from the lab into our lives

Information on the resveratrol study came from the article 'Compound found in berries and red wine can rejuvenate cells, suggests new study', by Richard Faragher, Lizzy Ostler and Lorna Harries, published in *The Conversation* on

14th November 2017. See: theconversation.com/compound-found-in-berries-and-red-wine-can-rejuvenate-cells-suggests-new-study-86945.

The quotations about the study came from the University of Exeter press release, 'Old human cells rejuvenated in breakthrough discovery on ageing'. See: www.exeter.ac.uk/news/featurednews/title_620529_en.html.

The figures from the Royal Pharmaceutical Society came from the article 'Drug development: the journey of a medicine from lab to shelf', by Ingrid Torjesen, published in *The Pharmaceutical Journal*, 12 May 2015. See: www.pharmaceutical-journal.com/publications/tomorrows-pharmacist/drug-development-the-journey-of-a-medicine-from-lab-to-shelf/20068196.article.

Similar information for the US came from 'The Drug Development and Approval Process', a report produced by FDAReview.org, a project of the Independent Institute. See: www.fdareview.org/03_drug_development.php.

Key sources of information on repurposing of drugs were the paper 'Drug repositioning: identifying and developing new uses for existing drugs', by Ted T. Ashburn and Karl B. Thor, published in *Nature Reviews* 3: 673–683 (2004). See: www.nature.com/articles/nrd1468.

And the article 'Repurposing Existing Drugs for New Indications', by Anna Azvolinsky, published in *The Scientist* on 1st January 2017. See: www.the-scientist.com/?articles.view/articleNo/47744/title/Repurposing-Existing-Drugs-for-New-Indications.

In addition to my own interview, a rich source of insights into the personal history, thinking and research preoccupations of Nir Barzilai – and background information about the TAME study – was the feature article 'The man who wants to beat back aging', by Stephen S. Hall, published in *Science* magazine on 16th September 2015. See: www.sciencemag.org/news/2015/09/feature-man-who-wants-beat-back-aging.

Information about the UK metformin study came from the original paper 'Can people with type 2 diabetes live longer than those without? A comparison of mortality in people initiated with metformin or sulphonylurea monotherapy and matched, non-diabetic controls', by C. A. Bannister et al., published in *Diabetes, Obesity and Metabolism* 16: 1165–1173 (2014). See: www.gwern.net/docs/longevity/2014-bannister.pdf.

Information about the TAME study was drawn from many sources, but key ones were the journal article 'Metformin as a tool to target aging', by Nir Barzilai, Jill P. Cradall, Stephen B. Kritchevsky and Mark A. Espeland, published in *Cell Metabolism* 23: 1060–1065 (2016). See: www.cell.com/cell-metabolism/pdf/S1550-4131(16)30229-7.pdf.

An interview with Nir Barzilai published in the March 2015 newsletter of the Healthspan Campaign, available at: www.healthspancampaign.org/2015/04/28/dr-nir-barzilai-on-the-tame-study.

And 'Metformin and the TAME Trial: Magic Pill or Monumental Tool?', a TEDMED blog written by Nir Barzilai on 29th August 2017. See: blog.tedmed.com/metformin-tame-trial-magic-pill-monumental-tool.

Acknowledgements

In reporting science, I am always aware of the fact there are many, many people besides the ones I get to speak to who have been involved in the research I am covering. I should like to acknowledge these legions, and to apologise to those whose contributions I have been unable to credit directly, or whose important work I have not even touched upon in this brief journey across an enormous field. Your published work has nevertheless informed and enriched my narrative considerably, for which I am extremely grateful.

I wish to express my sincere thanks to all the people I did speak to for this book and whose stories and insights have given me a whole new perspective on the journey through life, and much to ponder. In particular I should like to thank Richard Faragher for his readiness always to answer my queries, and for his encouragement, enthusiasm and irreverence (which gave me a much needed laugh from time to time); my dear friend Suzanne Cherney, a peerless editor, for her careful reading of my manuscript and judicious comments (though I didn't heed her advice to drop one or two lovely Scottish words!); my publisher and editor at Bloomsbury, Jim Martin and Anna MacDiarmid, and my agent Donald Winchester of Watson Little Ltd, for their warm support and Liz Drewitt, my copy-editor at Bloomsbury, for her meticulous attention to detail and good advice.

Many scientists were particularly generous with their time, expertise and readiness to share their stories and enthusiasms from the front line of research. They are, in alphabetical order: Peter Adams, Julie Anderson, Steve Austad, Mark Bagley, Nir Barzilai, Mark Blaxter, Dale Bredesen, Judy Campisi, Mar Carmena, Irina Conboy,

Mike Conboy, Lynne Cox, Caleb Finch, David Gems, John Hardy, Peter Hunt, Henri Jasper, Pankaj Kapahi, Brian Kennedy, Tom Kirkwood, Gordon Lithgow, Janet Lord, Janko Nikolich-Zugich, Linda Partridge, Emma Peat, Ram Rao, Wolf Reik, Martin Rossor and Thomas von Zglinicki. Kris Rebillot, Director of Communications at the Buck Institute, deserves a special mention also for her super-efficiency at organising my schedule during my visit to California, and her warm welcome.

Finally, I should like to express my special thanks to Mark Jones and his mother Pat Jones, and to John Jennings, whose personal stories of life in the shadows of serious illness are a poignant reminder of the relevance and urgency of the quest to understand the process of ageing and to find ways of preventing or treating its more distressing manifestations. Special thanks are due also to Dean Pomerleau, who has picked up and run with the science in his own everyday life, for sharing his experience of living on a very frugal diet.

Index